Copyright by

Don J. Easterbrook
Bellingham, Washington

All rights reserved. No part of this book may be reproduced in any form or by electronics without permission in writing from the author.

ISBN 9780692620748
Library of Congress Control Number 2016903398

Front cover: Mt. Baker steaming. Photo by D.J. Easterbrook

Mt. Baker (Photo by Lee Mann)

Mt. Baker from Three Tree Lake (Photo by Lee Mann)

TABLE OF CONTENTS

PAGE	
5	INTRODUCTION
9	**MT. BAKER ERUPTIONS**
9	EARLIEST ERUPTIONS (1.2–1.3 m.y.)
11	KULSHAN CALDERA (1.15 m.y.)
16	LAVA FLOW AND DIKES YOUNGER THAN KULSHAN CALDERA
22	OTHER LAVA FLOWS OUTSIDE KULSHAN CALDERA
26	BLACK BUTTES (375–290 kyr)
35	TABLE MT.
41	HEATHER MEADOWS
71	MT. BAKER VOLCANO
78	MT. BAKER SUMMIT CONE
85	SHERMAN CRATER
90	SCHREIBERS MEADOW CINDER CONE AND SULPHUR CR LAVA
102	VOLCANIC ASH ERUPTIONS
107	HISTORIC ERUPTIONS, STEAM ERUPTIONS
135	**GLACIATIONS OF MT. BAKER**
137	MT. BAKER DURING THE LAST ICE AGE
139	THE NOOKSACK ALPINE GLACIAL SYSTEM
147	GLACIATION OF HEATHER MEADOWS—ARTIST POINT
159	THE LITTLE ICE AGE
160	MODERN GLACIERS

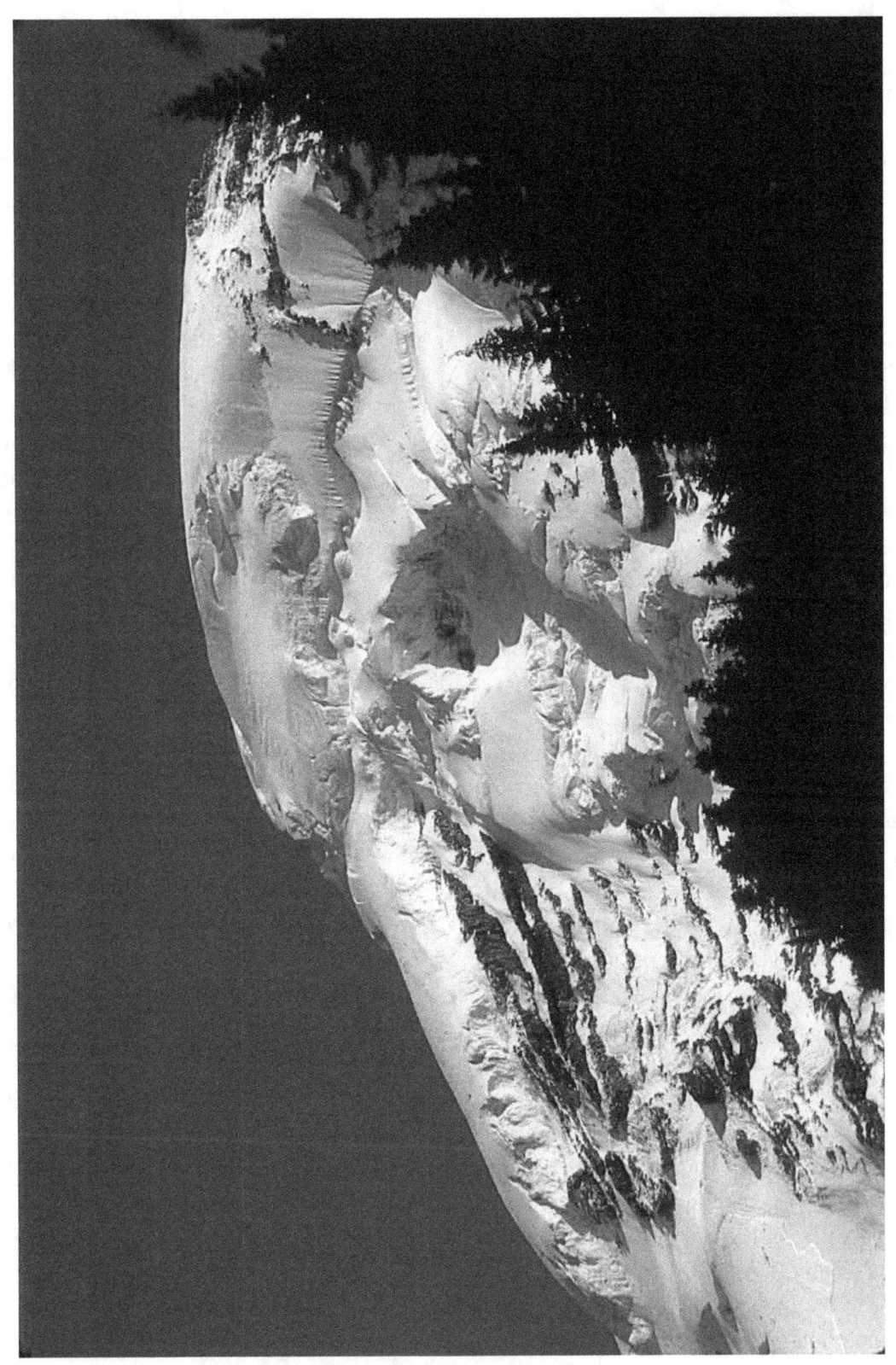

North face of Mt. Baker volcanic cone.

INTRODUCTION

My involvement with Mt. Baker began at an early age. From the time I was four years old (that was a long time ago!), my parents took me on long hikes in the mountains, and large parts of summers during my teenage years were spent hiking up the Coleman and Easton glaciers and to high alpine lakes with my brother, Bill. We would hike 10-15 miles into the high country and live on fish and berries for days when our packed food ran out. In my youth, I didn't know anything about geology, but that changed when I returned from the University of Washington with BS, MS, and PhD degrees in geology and I began seeing many of my childhood haunts in a new light. Since then, I have spent more than five decades conducting research on the geology of Mt. Baker. My first geologic experience on Mt. Baker began in 1959 when I became a faculty member at Western Washington University.

Over the many years, I've spent countless days hiking, camping, and climbing in the mountains around Mt. Baker and have always been more at home there than in cities. Lots of time there was spent with family and friends; professional geologists on field trips with regional, national, and international organizations; graduate students working on theses; field assistants; and undergraduate students. Many enjoyable evenings were spent around campfires discussing the geology of the area.

In the 1970s, I spent hundreds of hours photographing Mt. Baker and the North Cascades from the air with geologic colleague and friend Dave Rahm. Many times, we would spend entire days flying in the mountains, taking turns flying the plane and photographing geologic features. In recent years, I accumulated and studied thousands of air photos of glaciers on Mt. Baker. In addition to my own photos, Austin Post's remarkable collection of glacier photos taken every few years since 1950 were very useful in documenting advances and retreats of glaciers.

In more than 60 professional papers, I published the first isotope dates on Mt. Baker lavas, established the type localities of all volcanic ashes on Mt. Baker, radiocarbon dated all Mt. Baker volcanic ashes, radiocarbon dated glacial moraines dating back 12,000 years, and photographically documented advances and retreats of all Mt. Baker glaciers over the past few centuries.

History of the area

Mt. Baker is an active, glaciated volcano in the North Cascades of Washington about 30 miles due east of Bellingham (Fig. 1) and 15 miles south of the Canadian border. It is the highest peak (10,778 feet) in the North Cascade Range, the third-highest mountain in Washington, and is one of the snowiest places in the world. In 1999, the Mt Baker ski area set the world record for recorded snowfall in a single season (1,140 inches, 95 feet).

Indigenous Indian natives have known the mountain for thousands of years. They called it Koma Kulshan, meaning 'white sentinel.'

In 1790, a Spanish expedition under Manuel Quimper sailed from Nootka on Vancouver Island to explore the Strait of Juan de Fuca. During that six-week voyage, Gonzalo Lopez de Haro drew detailed charts and a sketch of the mountain, which the Spanish named "La Gran Montana del Carmelo."

Captain George Vancouver explored the area in the HMS Discovery and, while anchored in

Dungeness Bay on the Olympic Peninisula April 30, 1792, 3rd lieutenant Joseph Baker observed the volcano. Vancouver recorded in his journal:

"About this time a very high conspicuous craggy mountain ... presented itself, towering above the clouds: as low down as they allowed it to be visible it was covered with snow; and south of it, was a long ridge of very rugged snowy mountains, much less elevated, which seemed to stretch to a considerable distance ... the high distant land formed, as already observed, like detached islands, amongst which the lofty mountain, discovered in the afternoon by the third lieutenant, and in compliment to him called by me Mount Baker."

Six years later, the official narrative of Vancouver's voyage was published, including the first printed reference to Mt. Baker. By the mid–1850s, Mt. Baker was a well-known landmark to explorers, fur traders, miners and others who traveled in the region. In 1853, Isaac Stevens, the first governor of Washington Territory, wrote about Mt. Baker:

"Mount Baker ... is one of the loftiest and most conspicuous peaks of the northern Cascade Range; it is nearly as high as Mount Rainier, and like that mountain, its snow-covered pyramid has the form of a sugar-loaf. It is visible from all the water and islands ...and from the whole southeastern part of the Gulf of Georgia, and likewise from the eastern division of the Strait of Juan de Fuca. It is for this region a natural and important landmark."

Englishman Edmund Thomas Coleman, a veteran of the Alps, who lived in Victoria, British Columbia, made the first attempt to climb Mt. Baker in 1866. He picked a route via the Skagit River, but was turned back by local Native Americans who refused him passage. Later that same year, Coleman recruited Whatcom County settlers Edward Eldridge, John Bennett, and John Tennant to join him in his second attempt to climb the peak. They went up the North Fork of the Nooksack River, up the Coleman Glacier, and ascended to within several hundred feet of the summit but turned back in the face of an "overhanging cornice of ice" and threatening weather. Two years later, August 17, 1868, Coleman returned to the mountain with Eldridge, Tennant, and two new climbers, David Ogilvy and Thomas Stratton. They climbed to the summit via the Middle Fork of the Nooksack River, Marmot Ridge, Coleman Glacier, and the north margin of the Roman Wall.

Figure 1. Mt. Baker looking over Bellingham and Lake Whatcom.
The snowy peaks to the right are the Twin Sisters Range.

Previous work

The earliest study of Mt. Baker was made by Howard Coombs of the University of Washington in 1939. Since then, numerous Masters theses have been written about various aspects of the geology of the mountain by graduate students at Western Washington University and a few by graduate students at the University of Washington and Washington State University. The most comprehensive mapping of the geology of Mt. Baker was published by Haldreth, Fierstein, and Lanphere in 2003.

Geologic setting

Mount Baker is an active volcano and is part of the circum-Pacific 'ring of fire.' It is the youngest of several volcanic centers in the area and is one of the youngest volcanoes in the Cascade Range, less than ~100,000 years old. It has a thermally active crater that has erupted ash at least twice in the past 8,000 years.

The Mt. Baker volcanic area consists of four separate eruptive centers: (1) the Kulshan caldera, 1.15 million years old, (2) the Black Buttes, 300,000 to 500,000 years old, (3) the main summit cone, mostly less than ~100,000 years old, and (4) the Schreibers Meadow cinder cone, 8,800 ^{14}C years old. The main volcanic cone was built from successive eruptions of lava and volcanic breccia. Lava flows from the summit vent erupted between 30,000 and 10,000 years ago.

Figure 2. Topographic map of Mt. Baker.

MT. BAKER ERUPTIONS

Mt. Baker's volcanic cone (Fig. 2) is the predominant landform in the North Cascades, standing above non–volcanic peaks as a 'white sentinel' visible throughout the Puget Lowland. However, two older volcanic centers, the Black Buttes and Kulshan caldera, represent even larger eruptions, but their former edifices have collapsed or been eroded away. All of these eruptive sites are part of the larger Mt. Baker volcanic field, which includes.

(1) Pre–Kulshan caldera eruptions (1.3–1.15 million years ago).

(2) Eruption and collapse of Kulshan caldera (1.15 million years ago). The Kulshan caldera is a large, collapsed crater filled with volcanic ash and ashflows just northeast of Mt Baker's main summit cone. All of the once-present Kulshan volcanic cone is now gone as a result of collapse of the cone and subsequent erosion.

(3) Post–Kulshan caldera eruptions (1.15–0.5 million years ago). These are eruptions that occurred in the vicinity of Kulshan caldera but are younger than collapse of the Kulshan volcanic cone.

(4) Black Buttes eruptions (320–300,000 ago). A large volcanic cone was built just west of the present Mt Baker summit cone but has been deeply eroded by glaciers and is now known as the Black Buttes.

(5) Outlying eruptions (200–50,000 ago) occurred beyond the present volcanic cones.

(6) Mt. Baker volcano eruptions 45,000–10,000 years ago built the present summit cone of Mt Baker (Fig. 3).

(7) Schreibers Meadow cinder cone and Sulphur Creek lava flow (8,500 years ago) occurred at Schreibers Meadow on the south flank of Mt Baker.

(8) Eruptions of volcanic ash from Sherman Crater south of the main summit occurred during the past 6,000 years.

ERUPTIONS 1.3–1.15 MILLION YEARS AGO THAT PRECEDED KULSHAN CALDERA

Scattered remnants of lava flows and dikes older than the Kulshan caldera (1.15 million years) occur outside the caldera NE of the main Mt. Baker volcanic cone.

- Barometer Mtn. lava flow remnants between Wells and Anderson Creeks, 1.191 ± 0.094 million years old.
- Lava flow along the east wall of Deadhorse Creek 1.151 ± 0.016 million years old
- Dike between Swift Creek and Rainbow Creek, 1.180 ± 0.017 million years old.
- Dike on southern Cougar Divide, 1.293 ± 0.016 million years old.

Figure 3. Mt. Baker

KULSHAN CALDERA

Kulshan caldera (Figs. 4,5) is the remnant of a former volcano that collapsed following the largest eruption in the Mt. Baker volcanic field 1.15 ± 0.01 million years ago. The explosive eruption of more than 12 cubic miles of lava and ash was followed by collapse of the volcano inward and filling of the caldera with about 3,000 feet of massive ash flows and air fall ash (Figs. 4,6), overlain by well–bedded, laminated ash– rich sediments that accumulated in the caldera depression. Tributaries of upper Swift Creek have eroded deep gorges, exposing about 3,000 feet of the ashy sediments. Abundant pumice in the deposits contains about 10–15% crystals of plagioclase, pyroxene, hornblende, and biotite. Isotope dating of pumice from the caldera yielded an age of 1.149 ± 0.010 million years.

Figure 4. Ashy sediments in Kulshan caldera in upper Swift Creek between Table Mt. and Mt. Baker.

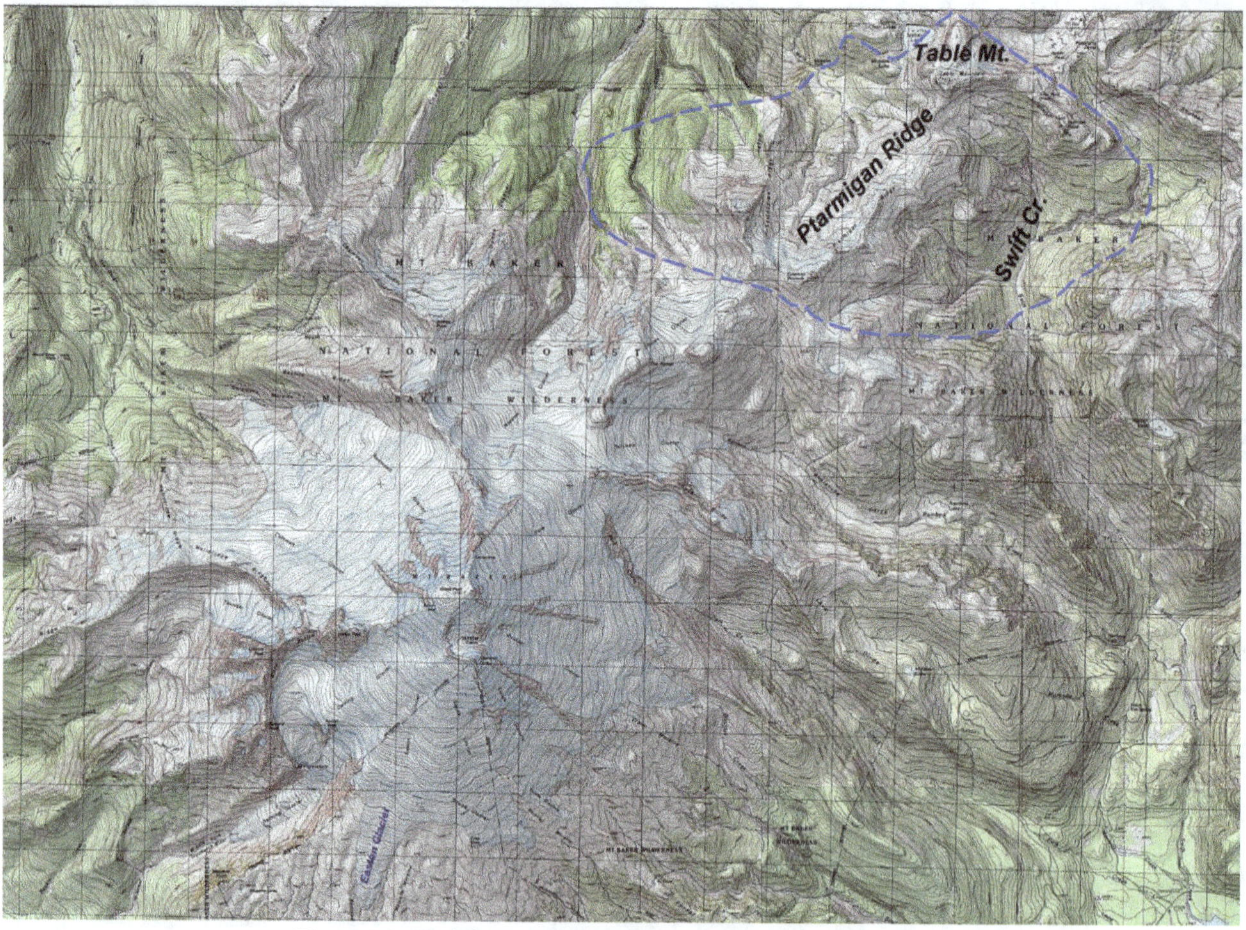

Figure 5. Kulshan caldera (blue line) NE of Mt. Baker.

Figure 6. Ash flows and air fall ash (white exposures) in Kulshan caldera. View from Artist Point looking toward Table Mt.

Several dikes, tabular bodies of lava intruded into fractures, and irregular masses of lava intrude the caldera ash flows in the upper drainage of Swift Creek (Fig. 6), one of which was dated at 1.155 ± 0.038 million years. Three lava domes in the middle of the caldera were dated at 1.127 ± 0.012 million years, 1.110 ± 0.012 million years, and 1.008 ± 0.007 million years.

A lava flow remnant overlying the ash flows along the north rim of the caldera was dated at 1.063 ± 0.012 million years and two lava flows in the western part of the caldera were dated at 1.052 ± 0.016 million years and 0.992 ± 0.014 million years.

No ash from this eruption has been found outside the caldera in the Cascades, most likely because of erosion by huge Cordilleran Ice Sheets that overran the area multiple times during the Ice Age. However, about a foot of volcanic ash from this eruption (Fig. 7) blanketed the region as far south as Seattle and along Hood Canal where it is known as the Lake Tapps ash. It also occurs as rounded pumice balls in stream sediments at the southern tip of Camano Island where it has been isotope dated at 1.12 ± 0.156 and 1.14 ± 0.184 million years old. Westgate, Easterbrook, Naeser, and Carson (1987) measured the age of the Lake Tapps tephra at 1.0 million years and determined the chemical composition of crystals and glass by electron microprobe, but these didn't match any known eruptions in the Cascade Range. However, the age and chemical composition of ash at Kulshan caldera are almost identical to the Lake Tapps tephra, so its source is now considered to be the Kulshan caldera.

Figure 7. Lake Tapps ash (white layer) in south Seattle (left) and Hood Canal (right). The chemical composition is identical to ash in Kulshan caldera and both are a million years old.

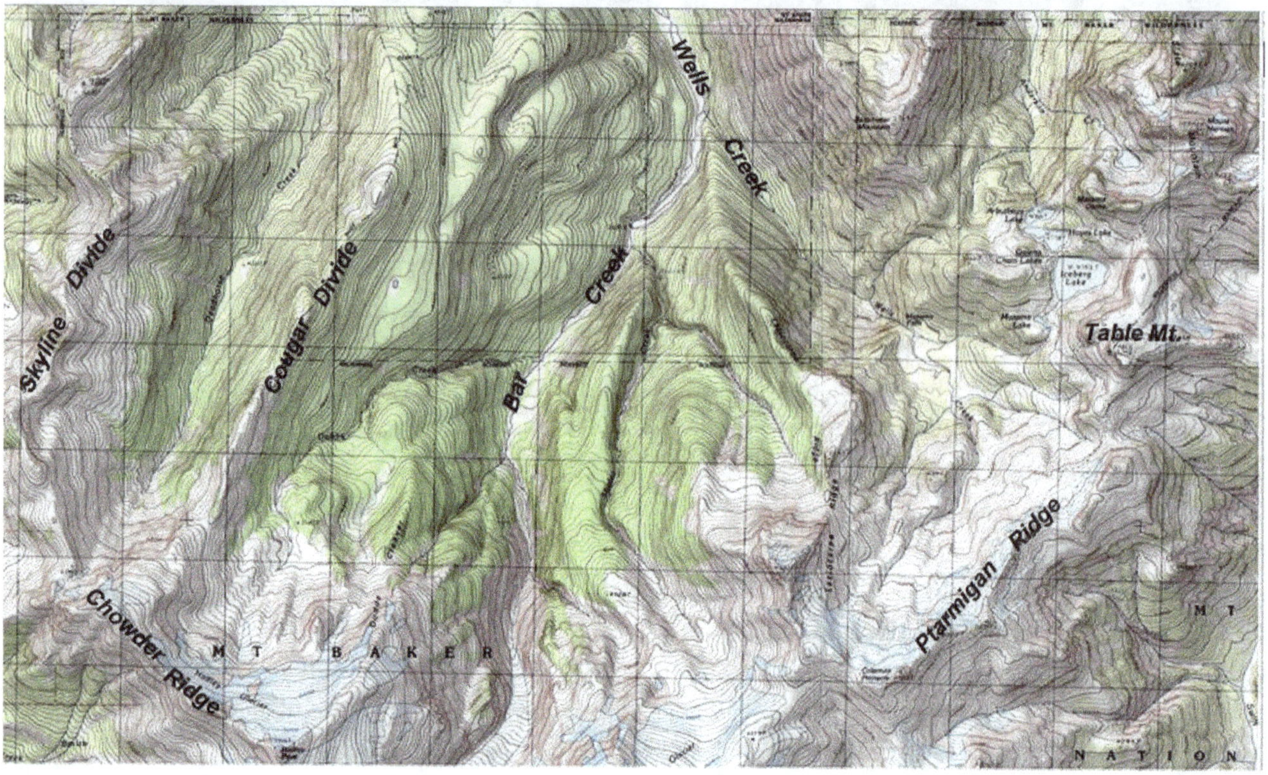

Figure 8. Topography NE of Mt. Baker between Cougar Divide and Table Mt.

LAVA FLOWS AND DIKES YOUNGER THAN KULSHAN CALDERA

Several post–caldera lava flows, dikes, and lava domes overlie or intrude the caldera ash flows both inside and outside the caldera margins. An ash flow inside the caldera in the basin of Swift Creek was dated at 1.155 ± 0.038 million years old. Three lava domes within the caldera were dated at:

      1.127 ± 0.012 million years
      1.110 ± 0.012 million years
      1.008 ± 0.007 million years

A remnant of a lava flow overlying ashflows at the rim of the caldera was dated at 1.063 ± 0.012 million years. Two lava flows within the caldera were dated 1.052 ± 0.016 and 0.992 ± 0.014 million years.

Dikes are tabular, intrusive bodies of igneous rock that often act as feeders to surface lava flows. Many dikes have intruded the ashflows of the Kulshan caldera and are thus younger than the ashflows that fill the caldera. They occur in all parts of the caldera. Most are 2-13 feet thick and dip 70°–90. A few pod-like intrusions are up to 160 feet thick.

LAVA FLOWS AND DIKES IN THE COUGAR DIVIDE–CHOWDER RIDGE AREA

At least six lava flow remnants and a number of dikes occur in the Cougar Divide area NE of Mt. Baker (Fig. 8). A 100–foot-thick lava flow north of Kulshan caldera was dated at 1.005 ± 0.017 million years and flow remnants on Cougar Divide were dated at 1.015 ± 0.018 and 1.052 ±0.016 million years.

A 230–foot thick lava flow near Chain Lakes was dated at 743,000 ± 34 years, a 400–foot-thick flow near Thompson Creek was dated at 878,000 ± 18,000 years, and a flow at Lookout Mt. was dated at 859,000 ± 14,000 years. A flow on the south wall of the Nooksack River canyon was dated at 202,000 ± 9,000 years.

Dozens of vertical dikes intrude rocks of the Nooksack Formation on Chowder Ridge and Dobbs Cleaver west of Kulshan caldera (Figs. 8, 9). Most are vertical, many are 3–26 feet thick, and a few are 60 feet thick. One dike was dated at 1.16 million years. A shallow intrusive dome on Cougar Divide was dated at 1.018 ± 0.008 million years (Figs. 10, 11).

Remnants of five lava flows occur along Cougar Divide. An extensive lava flow that caps the northern ridge of Cougar Divide is 600 feet thick and has been dated at 613,000 ± 8,000 years. This thick, ridge–capping lava flowed down an ancient stream valley whose sides have since been eroded away by streams and glaciers.

A 400–foot-thick lava flow along the east slope of Cougar Divide west side of Bar Creek has been dated at 334,000 ± 9,000 years (Figs. 10, 11). A small remnant of lava midway along the crest of Cougar Divide has been dated at 192,000 ± 8,000 years and a 400–foot-thick remnant of a lava along the west side of Bar Creek has been dated at 119,000 ± 23,000 years. The youngest lava flow along Bar Creek is a 200–foot-thick lava dated at 105,000 ± 8,000 years (Fig. 10).

Figure 9. Cougar Divide area NE of Mt Baker.

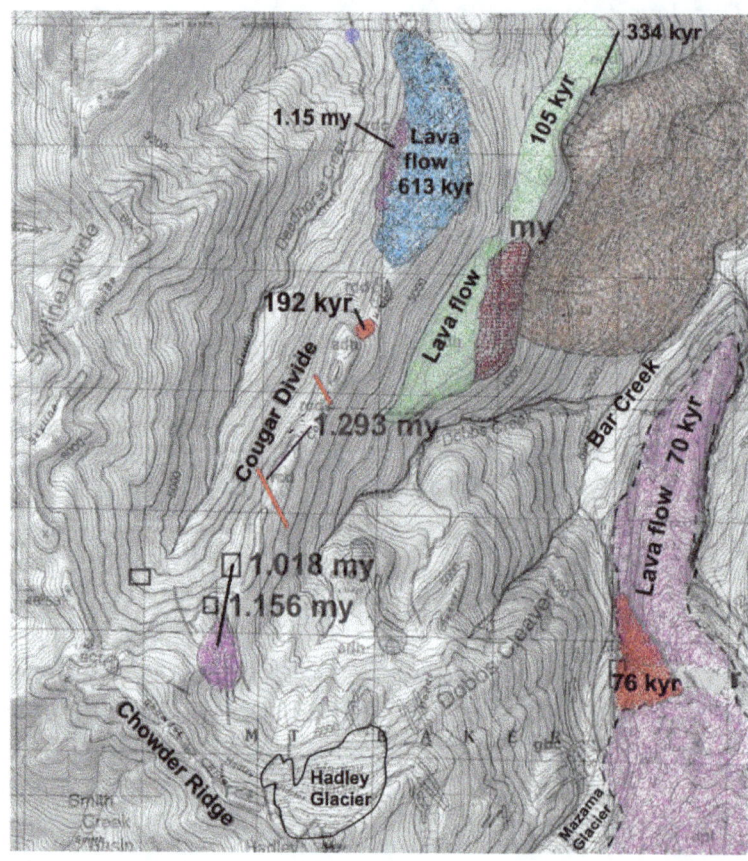

Figure 10. Remnant lava flows in the Cougar Divide area northeast of Mt. Baker. my=millions of years, kyr=thousands of years (Geologic data from Haldreth et al., 2003)

Figure 11. Cougar Divide.

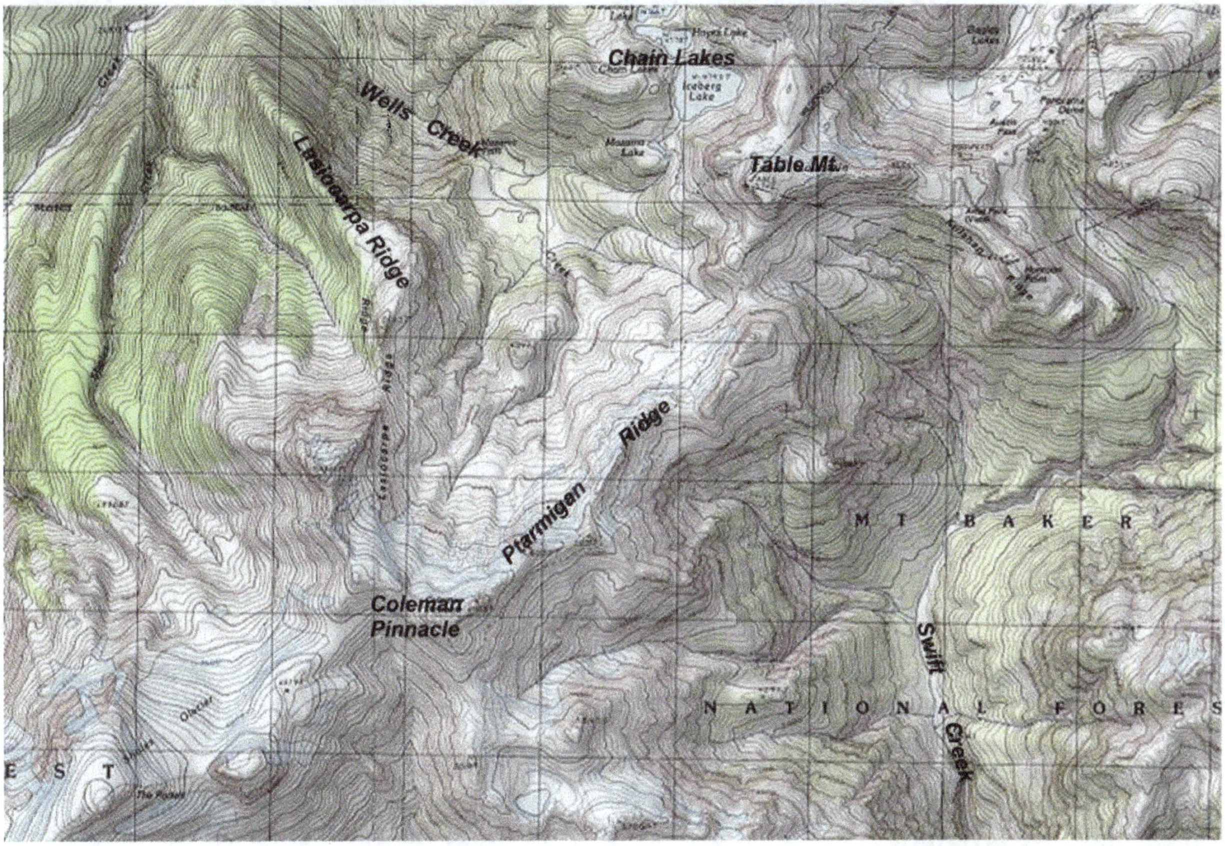

Figure 12. Topographic map of Lasiocarpa Ridge, Ptarmigan Ridge, Coleman Pinnacle, Table Mt., and upper Swift Creek.

Figure 13. Lasiocarpa Ridge lava flow.

OTHER LAVA FLOWS OUTSIDE KULSHAN CALDERA

A 100–foot–thick lava flow about half a mile outside the south rim of the caldera was dated at 1.005 ± 0.017 million years. Near Thompson Creek, a 400–foot–thick lava-flow was dated at 878,000 ± 18,000 years and a lava flow on Lookout Mountain was dated at 859,000 ± 14,000. A lava flow near Chain Lakes that forms a 1,500–foot–long ridge was dated at 743,000 ± 34,000 years (Fig. 10).

Lasiocarpa Ridge lava flow

Lasiocarpa ridge is a high, prominent ridge nearly two miles long (Figs. 12,13) made up of 300 feet of coarse flow breccia overlain by a massive 400–foot–thick lava flow dated at 515,000 ± 8,000 years. The lava thickens northwestward to 600–800 feet thick. The lava originally flowed down a deep valley, but the valley walls have now been completely eroded away, leaving the more resistant lava as a ridge. The flow straddles the rim of Kulshan caldera. At the upper part of the flow inside the caldera, the ridge is a high, steeply sloping pinnacle and basal breccia lies upon older lava, lake sediments, and hydrothermally altered ash flows. The lower part of the ridge beyond the caldera margin flattens out considerably and the lava lies directly on Mesozoic marine bedrock.

Ptarmigan Ridge and Coleman Pinnacle

Ptarmigan Ridge is a long linear ridge of lava between Table Mt. and Coleman Pinnacle. (Figs. 14, 15). It consists of (1) 100 feet of basal lava overlain by lava making the base of eastern Coleman Pinnacle, (2) a 200–foot high ridge of lava banked against the southwest side of Coleman Pinnacle; and (3) several dikes and pods that cut the lava of Coleman Pinnacle. Several dikes that are 15–60 feet thick and two that are 150–300 feet thick may have fed large eruptions lava that covered the floor of the caldera but have since been eroded.

A trail from the parking lot at Artist Point follows the base of Table Mt. to the divide separating the upper Swift Creek basin and upper Wells Creek Basin. The trail then splits, the right one leading into Chain Lakes at the base of Table Mt. and the left one traversing Ptarmigan Ridge to Coleman Pinnacle. Spectacular views of Mt. Baker and other alpine scenery may be seen from this trail.

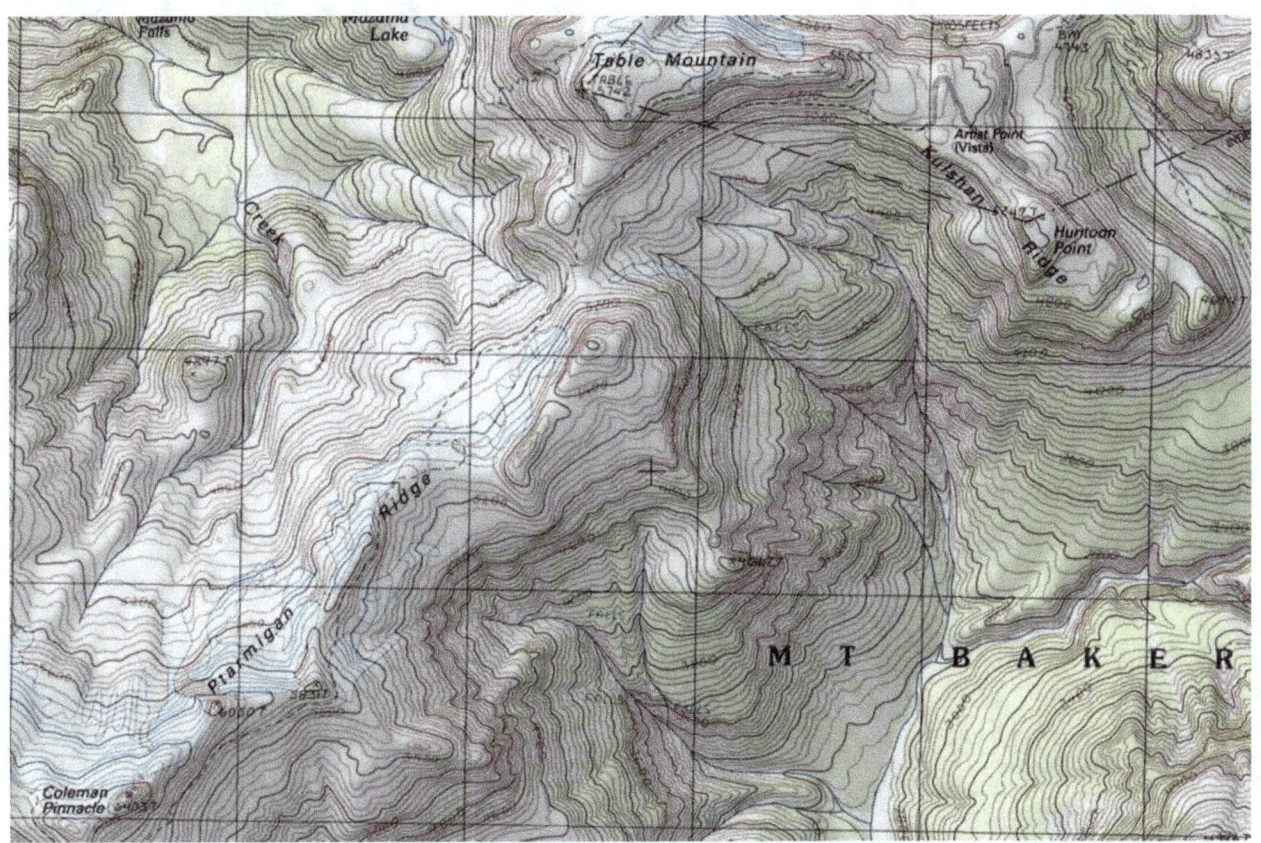

Figure 14. Topographic map of Coleman Pinnacle, Ptarmigan Ridge, and Table Mt.

Figure 15. Ptarmigan Ridge looking over Table Mt. to Mt. Baker.

Figure 16. Black Buttes (peaks on the right) and main summit cone (left).

BLACK BUTTES VOLCANIC CONE

The Black Buttes are the eroded remnants of an ancient volcano on the western flank of Mt. Baker (Figs. 16-25). They consist of two peaks, Lincoln Peak (9,096 ft.) (Figs. 19,20,22) and Colfax Peak (9,443 ft.) (Figs. 18,20), which have been deeply eroded by glaciers on their slopes. Remnants of lava flows exposed in the rock walls of the Buttes dip in opposite directions away from a former summit cone now long eroded (Fig. 16). Lava flows and volcanic breccia making up Colfax Peak dip eastward toward the present cone of Mt. Baker, whereas flows and volcanic breccia near Lincoln Peak dip in the opposite direction, indicating that the peaks are the remains of a volcanic cone whose central vent was between the two peaks. The lower part of the Black Buttes cone north of the Buttes has been eroded away, and to the east, lavas of the buttes disappear beneath the younger cone of Mt. Baker. Similar deep erosion characterizes lava flows hundreds of feet thick capping ridges high above present stream valleys between the Buttes and the Nooksack River.

Glacial erosion has played a significant role in the destruction of the Black Buttes cone, especially erosion in cirques at the heads of the Deming and Thunder glaciers. Exposed in the walls of the 2000–foot high cirque headwall of the Deming glacier are about 60 stratified volcanic breccia layers and thin lava flows, dipping 25°–30° away from the former central vent now occupied by the Deming glacier (Fig. 24). The total number of eruptive events is several hundred. Most are thin, brecciated lava flows.

About 30–35 lava flows are exposed on Heliotrope Ridge, the north wall of Thunder glacier cirque. The base of the west flank of the Black Buttes cone is exposed in the canyon of Wallace Creek, on Heliotrope Ridge, and at the snout of Deming Glacier. At Meadow Point on the south wall of the Nooksack Middle Fork above the Deming glacier, the base of a 1,300–foot–thick pile of Black Buttes lavas and breccias lies on metamorphic rocks.

At Bastille Ridge above the north wall of the Roosevelt glacier, 10–14 lava flows, each 30 to 200 feet thick, dip 15°–25° westward down the crest of the ridge and on the north side of Smith Creek. They apparently filled an ancient valley that once extended down Glacier Creek. The basal lava flow in Smith Creek basin was dated at 322,000 ± 12,000 years, and one of the uppermost flows capping Bastille Ridge was dated at 322,000 ± 9,000 years, within the general time frame of eruptions from the Black Buttes.

Remnants of lava flows at Cathedral Crag and Baker Pass above Schreibers Meadow on the south flank of Mt. Baker were contemporaneous with Black Buttes lavas. Each flow appears to be a single lava flow, about 400 feet thick. The flow at Cathedral Crag was dated at 331,000 ± 18,000 years and the flow at Baker Pass was dated at 333,000 ± 12,000 years.

Seven miles east of Nooksack Falls, a ridge–capping lava stands 700 feet above the floor of the modern canyon, making a cliff visible from the Mt. Baker Highway across the Nooksack River. The lava flow is about 300 feet thick with prominent columnar jointing (fractures) that formed during cooling and contraction of the lava. It flowed down an ancient valley but now caps a ridge because the lava was more resistant to erosion than the ancient valley sides, which have been eroded away, leaving the more resistant lava as a ridge top. Its topographic position suggests a possible correlation with the Table Mt. flows, but the isotope age of the flow is only 202,000 ± 9,000 years, 100,000 years younger than the Table Mt. flows. Near the junction of Wells Creek and the Nooksack River, a 1,300–foot long, 250–foot–thick lava flow remnant caps a ridge now 800 feet above river level. The flow was isotope dated in 1975 as 400,000 ± 100,000 years, but an isotope age of 149,000 ± 5,000 years was obtained more recently. A third flow remnant near Nooksack Falls 650 feet above the floor of Wells Creek has been isotope dated at 114,000 ± 9,000. The rate of erosion needed to invert these valley–filling lava flows to what are now ridge crests is extraordinary.

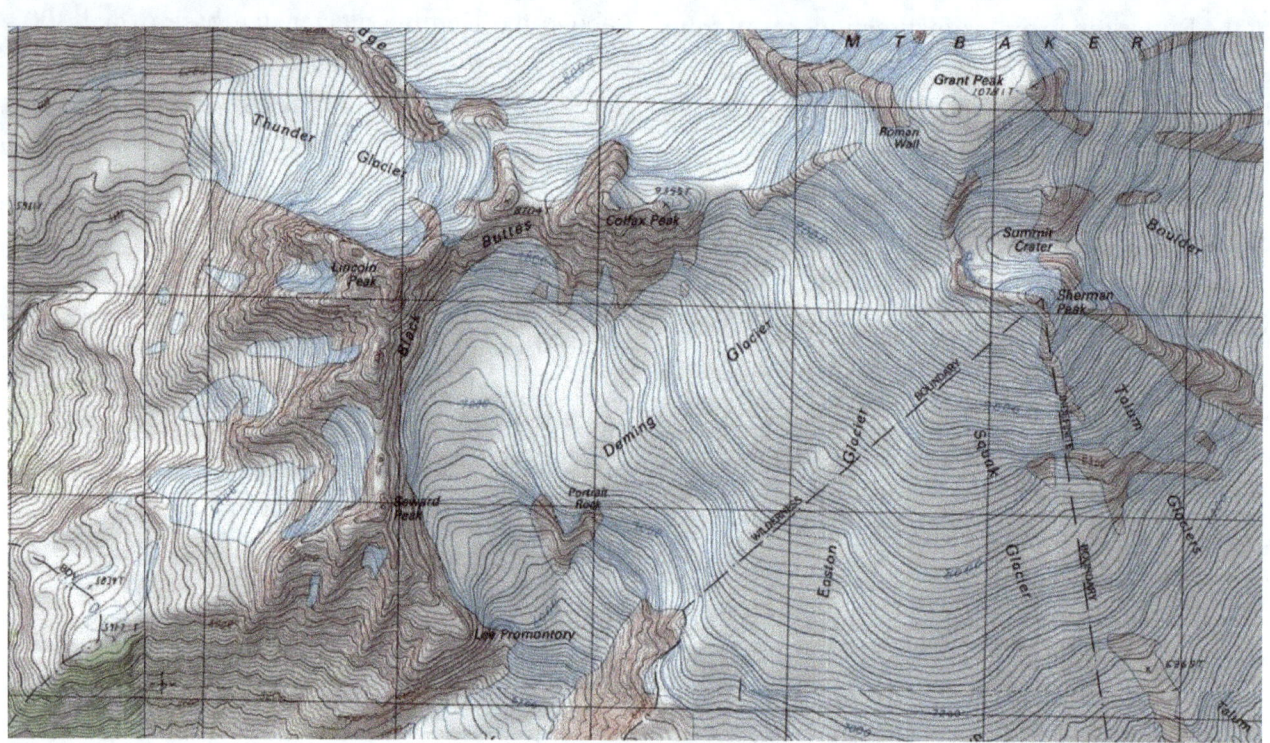

Figure 17. Topographic map of the Black Buttes, remnants of a former volcanic cone largely destroyed by glacial erosion.

Figure 18. Mt. Baker (left) and the eroded flanks of the older Black Butts volcanic cone (right). The dashed line shows the portion of the cone that has been eroded away.

Figure 19. The Black Buttes. Lavas making Mt. Colfax dip to the left and lavas making Mt. Lincoln dip to the right.

Figure 20. The Black Buttes. Colfax Peak is on the left and Lincoln Peak is on the right.

Figure 21. Lincoln Peak, the western remnant of the former Black Buttes volcanic cone.

Figure 22. Lincoln Peak, the western Black Butte. Lava flows and volcanic deposits of volcanic fragments are inclined 25-30° to the west (right) on the sides of the breached Black Buttes volcano.

Figure 23. Colfax Peak, Black Buttes.

Figure 24. [left] Colfax Peak, Black Buttes. [right] Lava flows and breccias making up Black Buttes above the Deming glacier..

TABLE MOUNTAIN

Table Mt. (Figs. 26, 27) is a high, flat–topped ridge of lava between Artist Point and Ptarmigan Ridge, standing about 1,500 feet above Bagley Lakes. A 300–400–foot–thick, ridge–capping flow rests on three lava flows at Heather Meadows and Panorama Dome. Two lava flows make up Kulshan Ridge, the lower of which is 500 feet thick and has slender, glassy, curving columns. South of Table Mt., the lowest lava flow is glassy, highly jointed, and its composition is different from lava at Coleman Pinnacle that caps much of Ptarmigan Ridge.

The 300-400–foot–thick lava flow capping the ridge at Table Mt. is a remnant of lava that originally flowed down an ancient stream valley whose sides have been completely eroded away (Fig. 25), leaving only the more resistant lava that once rested on the valley floor. The lava was probably erupted from the Black Buttes volcano on the west flank of the main summit cone of Mt. Baker. The summit of Table Mountain is about 1500 feet above Bagley Lakes, so the total amount of erosion must be 1,500 feet plus the height of the original valley side. The inverted topography at Table Mt. remains as testimony that a great deal of erosion has occurred during the past 300,000 years. The original valley sides, probably composed of ashy material having much less resistance to erosion than the lava flows in the valleys, eroded much faster than the lava, resulting in inversion of topography so that the flow originally occupying a valley, now makes up a resistant ridge (Fig. 27-29).

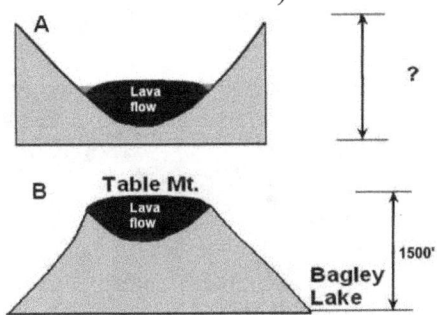

Figure 25 Topographic inversion of Table Mt. lava flow. (A) Emplacement of Table Mt. lava. (B) Erosion of less resistant valley sides, leaving the lava flow as a ridge crest.

The first isotope dates of lava flows in the Mt. Baker region came from lava near the base of Table Mt, which was dated at 400,000 ± 200,000 years, based on the average of three analyses (Easterbrook, 1975). In 2003, the U.S. Geological Survey dated the top flow on Table Mt. at 309,000± 13,000 years and a basal lava flow at Heather Meadows at 301,000 ± 8,000 years. Lava beneath Coleman Pinnacle has been dated at 306,000 ± 13,000 years.

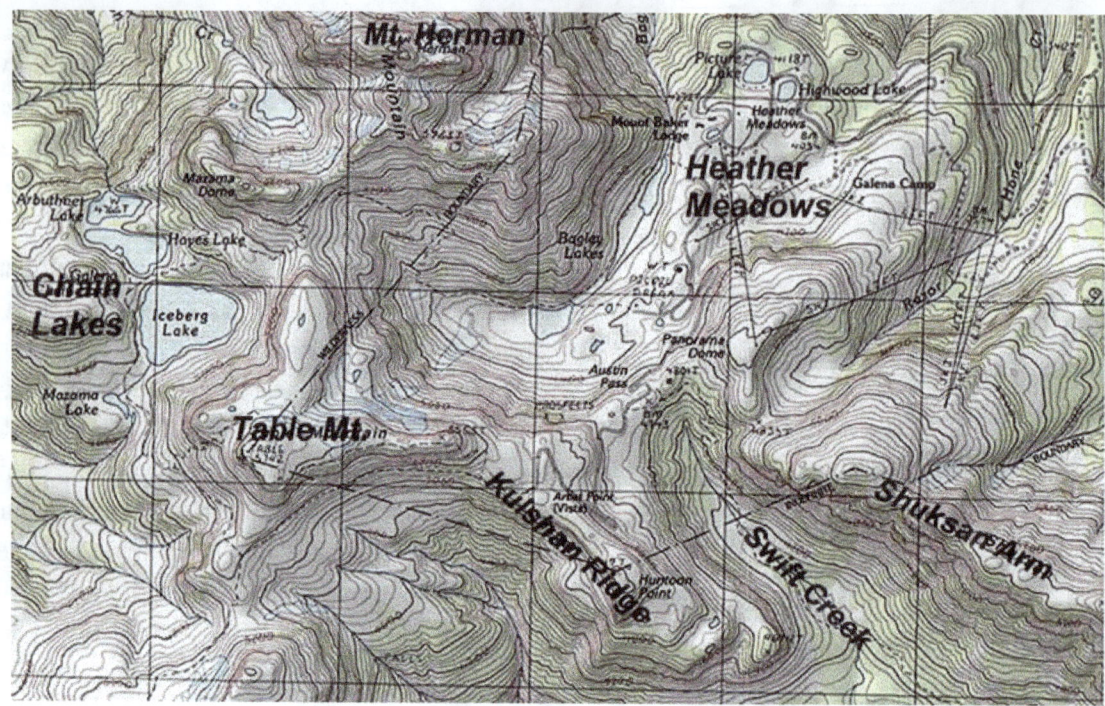

Figure 26. Topographic map of Table Mt.

Figure 27. Table Mt. looking west. The linear ridge of Table Mt. marks the floor of a former stream valley whose sides have been eroded away. The trail across the middle of Table Mt. is the Ptarmigan Ridge–Chain Lakes trail from Artist Point.

Figure 28. Table Mt. viewed from Austin Pass.

Figure 29. Looking over Table Mt. to Mt. Shuksan.

Figure 30. Bagley Lake at the head of the cirque below Table Mt.

HEATHER MEADOWS

The lava flows making up Table Mt. extend eastward where they cover Heather Meadows (Figs. 31-35). Numerous road cuts expose column jointed lava flows. As lava cools, it contracts and forms six-sided fractures that pull away from one another, forming columns (Fig. 31). The long axis of each column forms at right angles to the cooling surface of the flow, generally (vertical), so as long as the flow of lava is uniform, the columns will be upright. However, if the cooling surface is disturbed, as when lava flows over irregular topography, the columns will become distorted (Fig. 32).

A lava flow along the highway at Heather Meadows has been isotope dated at 301,000 ± 8,000 years, typical of the age of the lava making up Table Mt.

Figure 31. Columnar jointing in a Mt. Baker lava flow along the highway below Heather Meadows.

Figure 32. Swirling columnar jointing in a lava flow along the road from Austin Pass to Artist Point. The columns are distorted because the lava flowed over a steep slope that disturbed the normal temperature gradient within the flow, which controls the formation of columns.

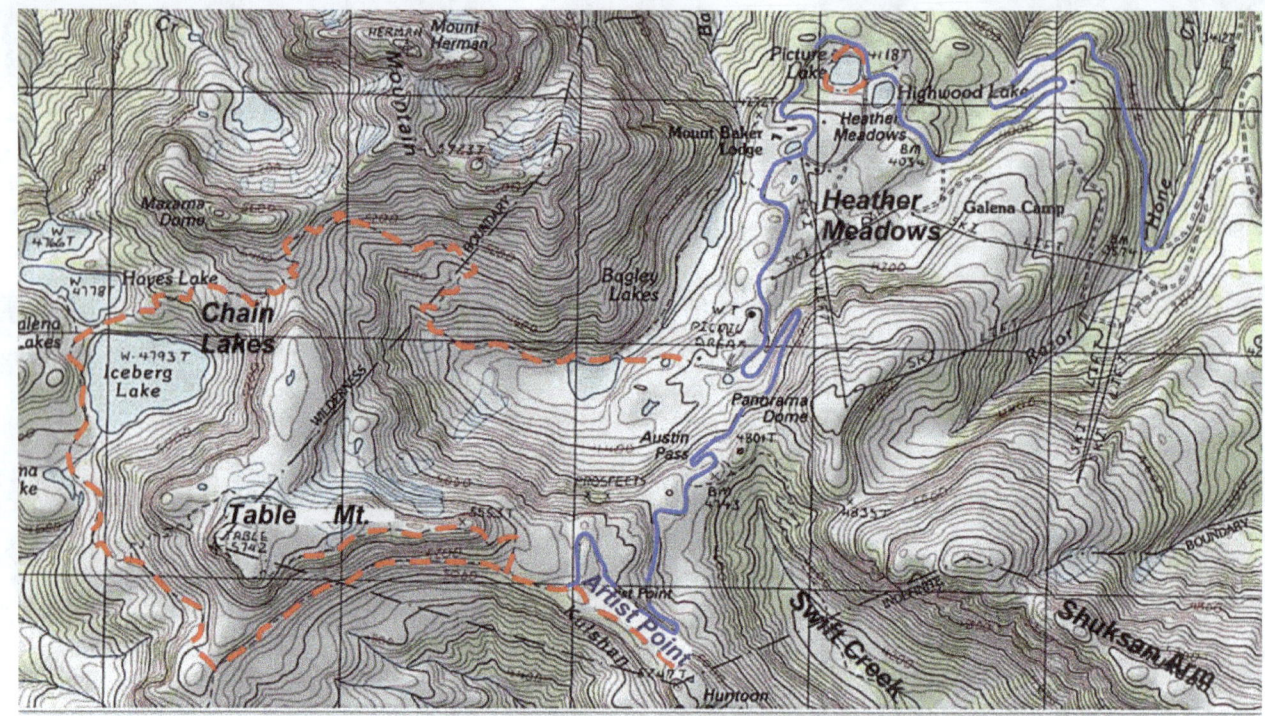

Figure 33. Heather Meadows at the end of the Mt. Baker Highway (blue line). The red dashed lines are the Chain Lakes and Table Mt. trails.

Figure 34. Heather Meadows and vicinity.

Picture Lake (Figs. 36-38) is famous for its spectacular vista of Mt. Shuksan, which is often reflected in the surface of the lake. From the parking area on the left side of the road at the edge of the lake, walk directly across the road. The ridge on the far side of the road is a glacial moraine that holds in Picture Lake. It was made by an alpine glacier that formerly occupied this part of Heather Meadows. The white, ashy layer exposed in the road cut (Fig. 40) is Mazama ash from a cataclysmic eruption at Crater Lake in Oregon 6,800 years ago. Thus, the glacial sediments upon which it rests must be older.

Walk along the path at the edge of the lake. When the lake is calm, Mt. Shuksan is reflected in it. Mt. Shuksan is made up of greenschist, a former basaltic lava that has been recrystallized by low heat and high pressure far beneath the Earth's surface. It has been pushed over younger rocks, uplifted thousands of feet, and the former overlying rocks eroded away by glaciers and streams.

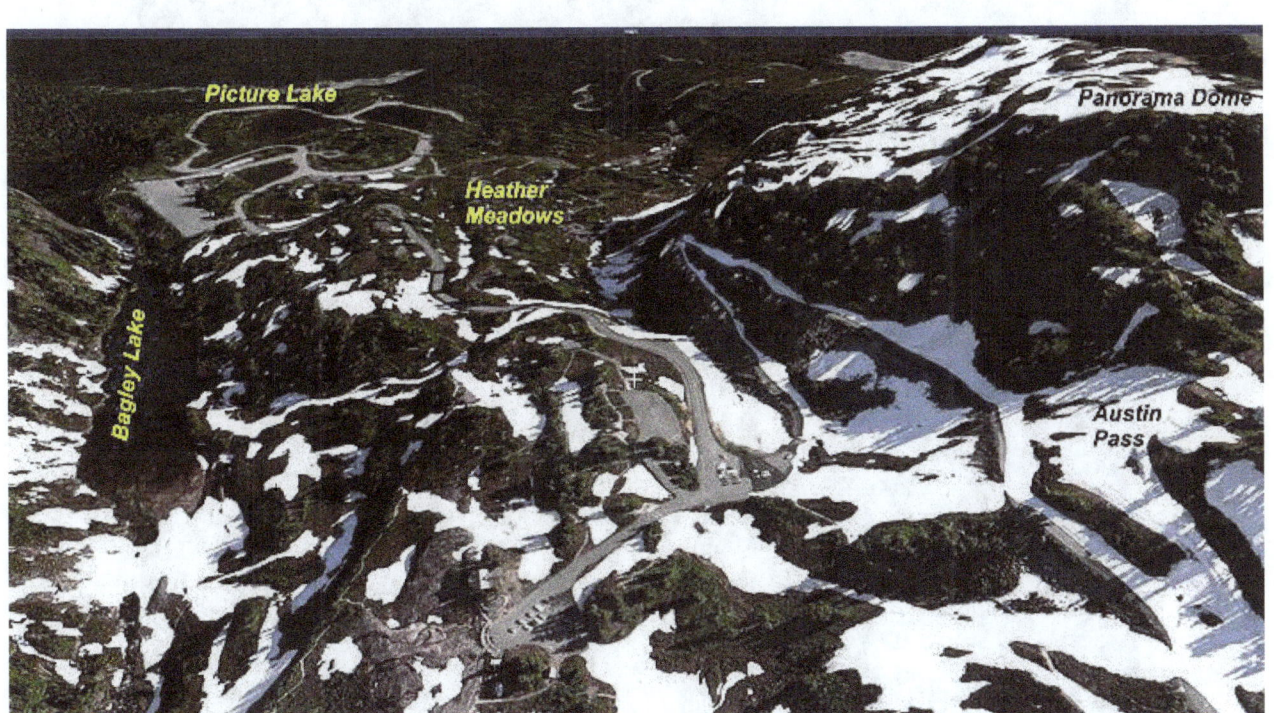

Figure 35. Heather Meadows and vicinity.

Figure 36. Mt. Shuksan reflected in Picture Lake at Heather Meadows.

Figure 37. Mt. Shuksan, Picture Lake.

Figure 38. Heather Meadows and Mt. Shuksan in winter.

The path around the lake leads around to the far side of the lake where a small peninsula extends out into the lake (Fig. 39). The material making up the peninsula is the Cathedral Crag ash, a coarse, sandy, black ash erupted from Mt. Baker 5800 years ago. The ash is fairly cohesive and thus somewhat less erodible. A few inches of it can be seen in low edges exposed at the lake shore.

Figure 39. Grassy peninsula composed of Cathedral Crag ash extending into Picture Lake.

Figure 40. Mazama ash from Crater Lake, Oregon, on glacial deposits exposed in roadcuts across the road from Picture Lake.

TABLE MT. TRAIL

The Table Mt. trail begins at the Artist Point parking lot at the end of the road Figs. 41, 42). The trail is short, but steep, rising about 500 feet above the parking lot, with steep drop–offs beside the trail. Appropriate footwear is strongly recommended. Good exposures of the ridge–capping lava flow occur along the trail. The ridge crest is fairly flat, rising from 5,553 at the east end to 5,742 at the west end. The headwall of Bagley Lakes cirque drops precipitously to the north into upper Bagley Lake. A boulder of old volcanic breccia that probably came from Mt. Herman across the valley to the north lies at the east end of the ridge. The only way to get this boulder across the intervening valley is transportation by a glacier, in this case the Cordilleran Ice Sheet 15-20,000 years ago. That means that the top of the ice sheet here had to be about 6,000 feet above sea level.

Figure 41. Table Mt. from Artist Point.

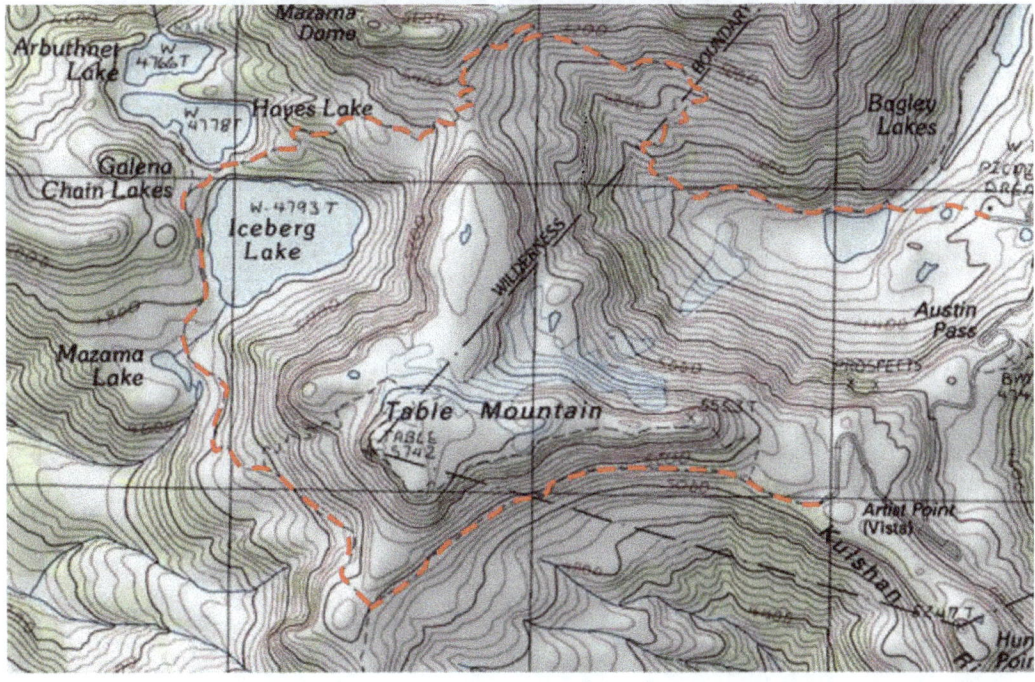

Figure 42. Topographic map of Table Mt. and Chain Lakes loop trail (red).

CHAIN LAKES TRAIL

The Chain Lakes trail begins at the Artist Point parking lot and traverses the slope to the west below Table Mt. (Fig 42). At the west end of Table Mt. the trail splits, one path continuing westward to Coleman Pinnacle and Camp Kiser and the other turning to the north around the west end of Table Mt. The Chain Lakes trail drops from 5200 feet at the trail junction to about 4800 feet at Iceberg Lake (Figs. 44–45). Along the trail between Ptarmigan Ridge and Iceberg Lake, Mazama ash (6800 years old) and Rocky Creek ash (5800 years old) are exposed (Fig. 43).

Chain Lakes consist of three lakes, Iceberg, Hayes, and Arbuthnet Lakes, all of which are tarns (glacial lakes formed in cirque basins). One cirque basin is occupied by Iceberg Lake and Hayes, and Arbuthnet Lakes share a second basin with a ridge across the lower end.. At one time, probably about 10,000 years ago during a cold period known as the Younger Dryas, an alpine glacier filled these cirques and glacially eroded the basins. The glacier receded about halfway upvalley where glacial erosion hollowed out the basin now occupied by Iceberg Lake.

Figure 43. Mazama ash and Rocky Creek ash exposed along the trail to Chain Lakes. Mazama ash is 6,800 ^{14}C years old and Rock Creek ash is 5,800 ^{14}C years old.

Figure 44. Table Mt. overlooking Chain Lakes (Iceberg and Hayes Lakes). Note the thick, ridge–capping lava flow of Table Mt.

Figure 45. Iceberg Lake, a glacial tarn eroded by an alpine glacier, probably about 10,000 years ago.

Figure 46. Large rock slide on Mt. Herman above upper Bagley Lake.

The trail continues around the west shore of Iceberg Lake to a junction with a trail that crosses the glacially–created ridge separating Iceberg and Hayes Lakes and drops down to Hayes Lake (Fig. 42). The main trail ascends 600 feet to the pass north of Iceberg Lake (Fig. 42) where views of Mt. Shuksan and Mt. Baker are spectacular. The trail then drops about 1000 feet down the Bagley Lakes cirque headwall at the head of Bagley Creek. About midway along the trail from the pass to upper Bagley Lake, the path encounters a large rockslide (Fig. 46).

At a trail junction at the east end of upper Bagley Lake, the right hand fork crosses glacially scoured lava flows leading to the parking lot at the Austin Pass ranger station and the left hand fork continues for about a mile to Mt. Baker Lodge.

Ptarmigan Ridge trail

The Ptarmigan Ridge trail begins at the junction with the Chain Lakes trail at the west end of Table Mt. (Fig. 48) It traverses the upper part of Wells Creek drainage, passing several small glacial moraines before crossing a ridge and traversing the upper slopes of the Swift Creek drainage just below the ridge crest. Be wary crossing a perennial snowfield (just below the peak labeled 5831 on Fig. 48) that is quite steep and can be dangerous for people without proper hiking boots. The trail ends on the west side of Coleman Pinnacle. Coleman Pinnacle (Figs. 47,48) consists of lava erupted from the Black Buttes, isotope dated at 305,000 ± 6,000 years.

Figure 47. Mt. Baker with Ptarmigan Ridge in foreground. The dark peak on the far left is Coleman Pinnacle .

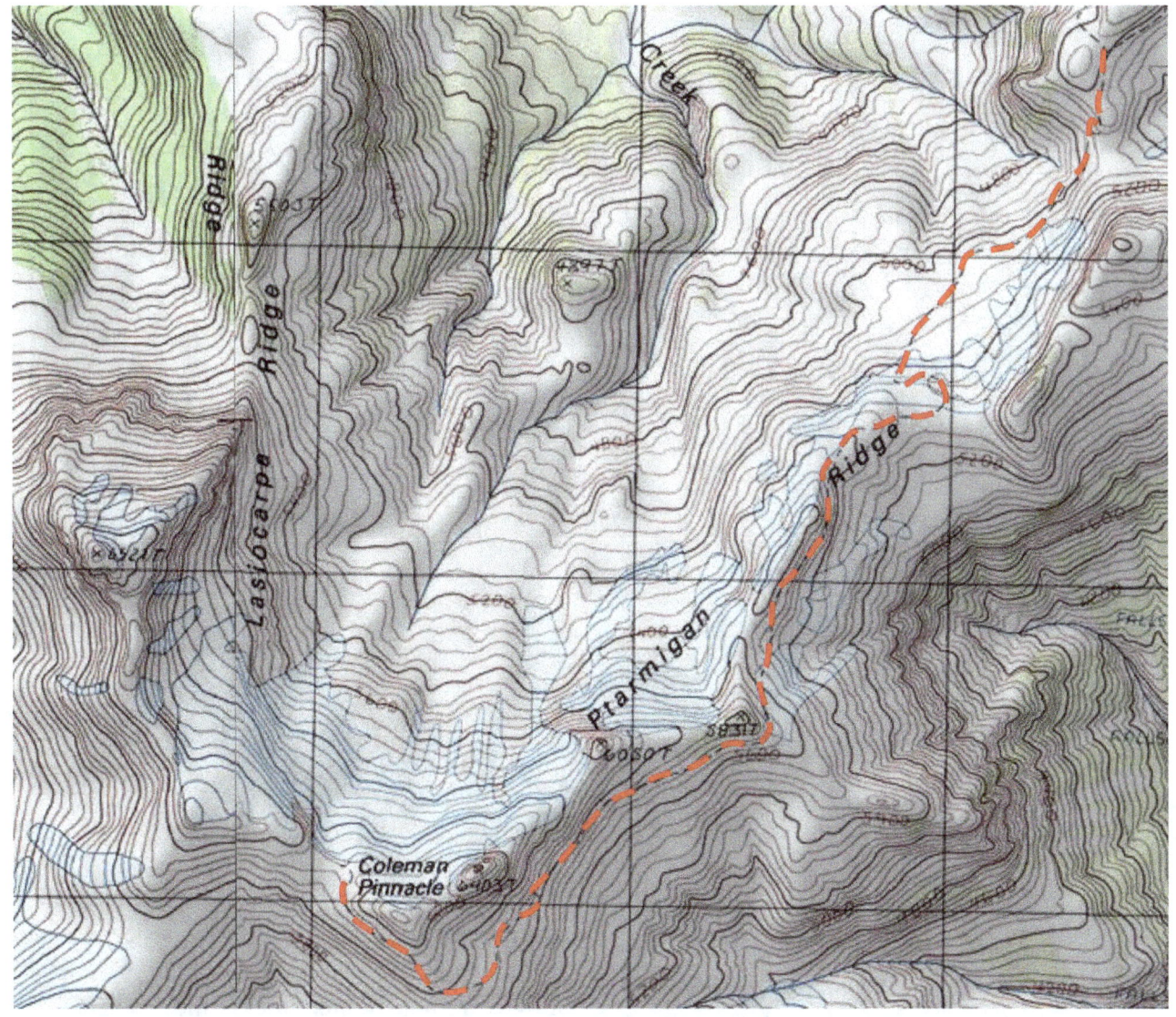

Figure 48. Topographic map of Ptarmigan Ridge, Coleman Pinnacle, and Ptarmigan Ridge trail (red).

Artist Point trail

The Artist Point trail begins at the parking lot and extends eastward along Kulshan Ridge (Fig. 42). At the beginning of the trail, the surface of a lava flow has been extensively grooved and polished (Fig. 49) by a glacier that flowed southward cross the ridge at right angles to the crest. The grooves are all oriented north–south parallel to the direction of the flow of the Cordilleran Ice Sheet that passed over the area about 15–20,000 years ago. Good examples of the glacially polished and grooved rock occur all along the ridge.

Figure 49. Glacially grooved and polished rock near Artist Point parking lot.

Exceptionally scenic views of both Mt. Shuksan and Mt. Baker are afforded all along the ridge. The view of Mt. Baker is across the upper drainage of Swift Creek. Light-colored deposits exposed in the valley walls consist of volcanic ash flows and sediments within the Kulshan caldera. The ridge above is Ptarmigan Ridge.

Shuksan Arm, the long ridge on the left that extends eastward from Panorama Dome to Mt. Shuksan, is composed of an unusually young granite known as the Lake Ann stock (a stock is an irregular intrusion of molten rock) (Fig. 50). It is particularly noteworthy because of its very young age, 2.7 million years, for a deep-seated granitic body of this size. The granite crystallized slowly from a molten rock miles beneath the Earth's surface, was later uplifted by mountain-building forces, and exposed by erosion of overlying rock, all within the past 2.7 million years. The top of the Lake Ann granite intrusion has a deep orange color due to oxidation of the rocks (Figs. 50, 51). This typically occurs later in crystallization of granitic intrusions due to alteration of the rocks by residual hot solutions.

Figure 50. Mt. Shuksan and Shuksan Arm (ridge left of Shuksan) from Artist Point.

Figure 51. Shuksan Arm viewed from Artist Point. Red-orange colored rocks are oxidized Lake Ann granite intrusion,

Lake Ann trail

The Lake Ann trail (Fig. 52) is an 8–mile hike (roundtrip) with lots of up and down, but the scenic reward at the end of the trail is spectacular. The trail begins at Austin Pass, drops fairly steeply 800 feet down into the valley of upper Swift Creek, continues on a gentle slope for a bit less than two miles and then climbs up about 1000 feet to a saddle above Lake Ann. The close-up view of Mt. Shuksan and its glaciers from here is truly remarkable (Fig. 58). Figures 53-57 are aerial views of Mt. Shuksan. From Lake Ann the trail continues on toward the lower Curtis glacier where large masses of ice break off the upper Curtis glacier and thunder down near–vertical cliffs to the glacier below.

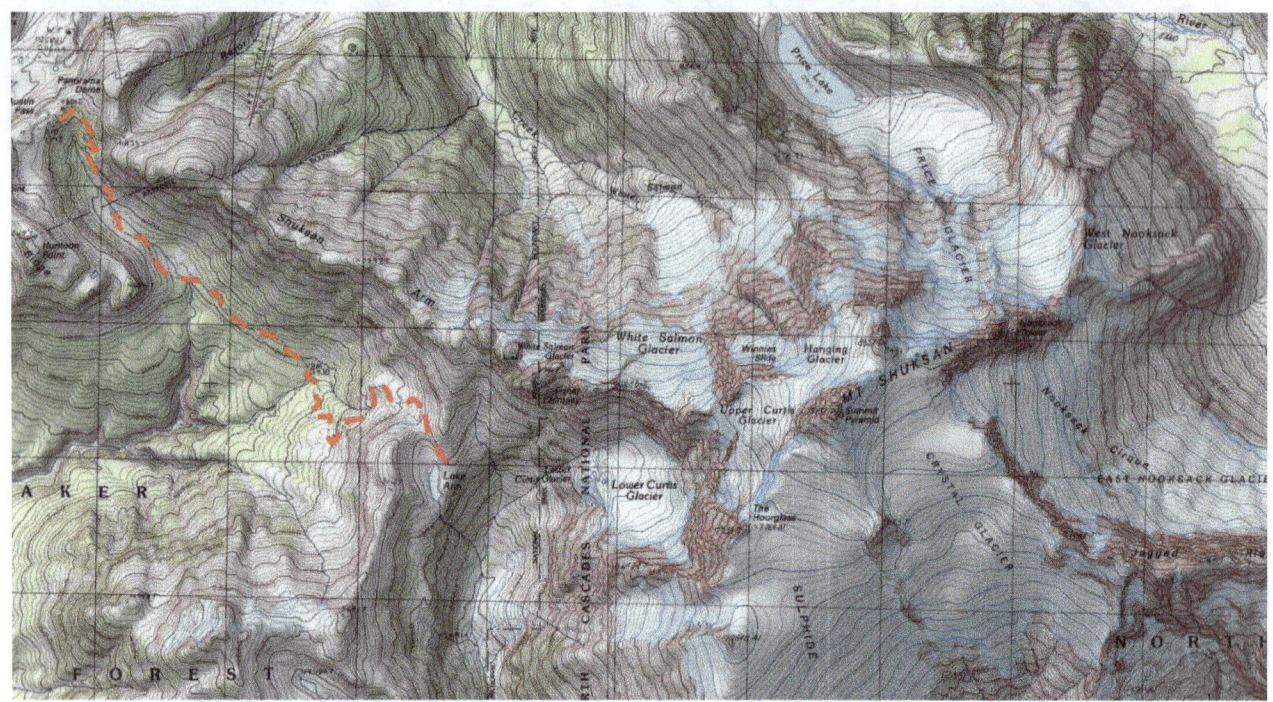

Figure 52. Topographic map of Mt. Shuksan and the Lake Ann trail (red).

Figure 53. Looking over Sherman Crater (left) on Mt. Baker to Mt. Shuksan (background).

Figure 54. Mt. Shuksan from above Artist Point.

Figure 55. Mt. Shuksan from above Lake Ann.

Figure 56. Mt. Shuksan summit.

Figure 57. Mt. Shuksan.

Figure 58. (left) Upper and lower Curtis glaciers on Mt. Shuksan viewed from Lake Ann. (right) Terminus of the lower Curtis glacier. The dark bands in the ice are annual debris layers bounding layers of snow and ice.

Mt. Shuksan is composed of mostly of greenschist and phyllite that has been thrust faulted over Paleozoic marine sedimentary rocks (Figs. 59, 60). The greenschist was formed by recrystallization of former basaltic lava by low temperature and high pressure metamorphism. The phyllite has been recrystallized from shale by low temperature, high pressure metamorphism.

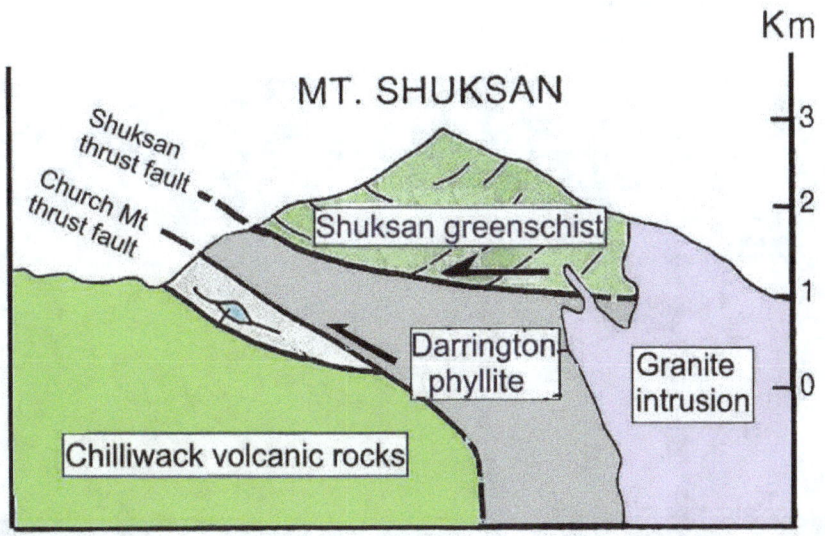

Figure 59. Geologic cross–section through Mt. Shuksan. Metamorphic rocks (greenschist and phyllite) have been thrust faulted over Paleozoic marine sedimentary rocks and intruded by the Lake Ann granite.

Figure 60. Greenschist of Mt. Shuksan thrust faulted over Paleozoic marine sedimentary rocks.

MOUNT BAKER VOLCANO

The principal Mt. Baker volcanic cone (Figs. 61, 62) rises 10,781 feet above sea level. It is made of alternating lava flows, ash flows, and air–fall ash. Most of the cone consists of lava flows, with smaller amounts of volcanic breccia, erupted within the past 90,000 years. Most of the cone is younger than about 40,000 years and the upper part of the cone is younger than 20,000 years.

Figure 61. Mt. Baker eruptive centers–Black Buttes (left), Grant Peak summit cone (center), and Sherman Crater (right of the summit). (Photo by Lee Mann)

Figure 62. Lava flows dipping from the Mt. Baker summit cone.

Figure 63. Cone–building lava flows on the flanks of Mt. Baker. Black lines delineate lava flows. Numbers are ages of the lavas in thousands of years.

Cone–building lava and breccia

The Mt. Baker volcanic cone may include as many as 200 lava flows, mostly dipping 25°–35,° that flowed radially away from the central cone (Fig. 63). The oldest lavas beneath the Mt. Baker cone are 650 feet of flows downstream from the terminus of the Park Glacier (Fig. 64), dated at 140,000 ± 55,000 years. About 1,500 feet of overlying lava flows, dipping 10°–15°E away from the cone, are exposed (1) in the cliff above the terminus of Park glacier (Fig. 64); (2) the surface of Boulder Ridge between the Boulder and Park Creeks (Fig. 65); (3) the upper basin of Boulder Creek below the snout of Boulder Glacier; and (4) the lower end of Boulder Cleaver (Fig. 63). The rim–forming flow on the cliff above Park Creek has been dated at 80,000 ± 14,000 years and a flow at the terminus of the Boulder Glacier has been dated at 90,000 ± 52,000.

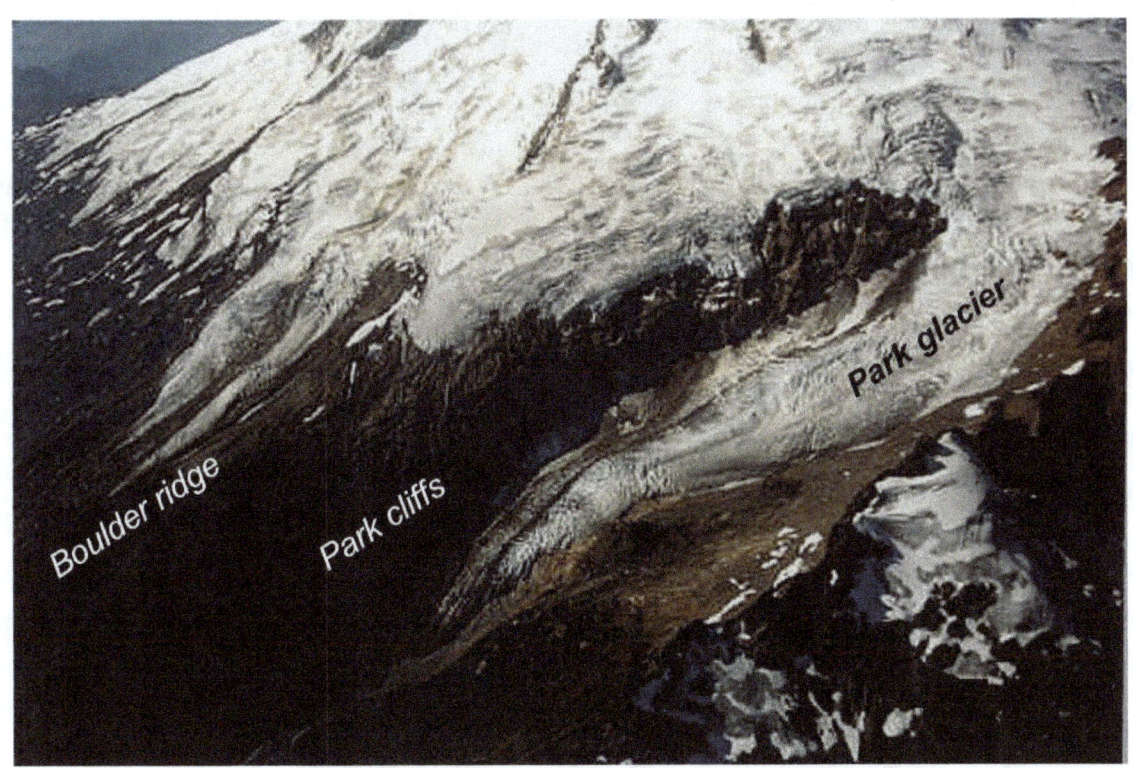

Figure 64. Valley of Park Glacier. 140,000–year–old lava is just downvalley from the terminus of the Park Glacier. The lava ridge–capping flow on Boulder ridge above the valley is 80,000 years old. (Photo by Austin Post)

Figure 65. Boulder Ridge lava flow between Boulder Creek and Park Creek.

Lava flows near The Portals on the east side of the Mt. Baker cone (Fig. 66) have been dated at 70,000 ± 7,000 years and flows on the east rim of Bar Creek gorge have been dated at 76,000 ± 7,000 years.

In the gorge of Ridley Creek, a single thick lava flow lying on metamorphic basement has been dated at 43,000 ± 5,000 years (Fig. 67). Lava at the distal edge of the Crag View fan about half a mile north of Schreibers Meadow on the floor of Sulphur Creek (Fig. 67) has been dated at 36,000 ± 14,000 years. A thick lava flow about half a mile south of Baker Pass (Fig. 67) has been dated at 32,000 ± 14,000 years.

A flow near the top of the Crag View lava fan has been dated at 11,000 ± 9,000 years and a flow between Easton and Deming glaciers has been dated at 9,000 ± 11,000 years (Fig. 67).

A 200–foot thick lava flow exposed beneath the Roosevelt and Coleman glaciers has been dated at 24,000 ± 16,000 years and thick lava that flowed down Glacier Creek below the Coleman Glacier has been dated at 14,000 ± 9,000 years. A 200–foot thick lava flow that caps much of the ridge between the lower Roosevelt and Coleman glaciers has been dated at 9,000 ± 7,000 years.

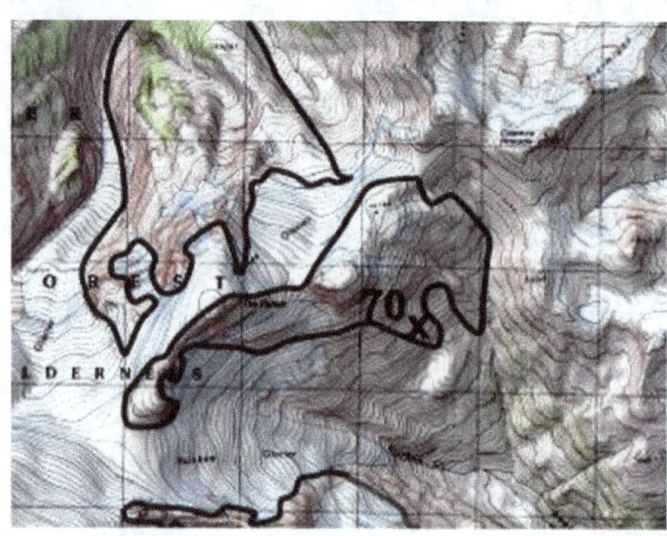

Figure 66. Lava flows from Mt. Baker cone near The Portals dated at 70,000 to 76,000 years.

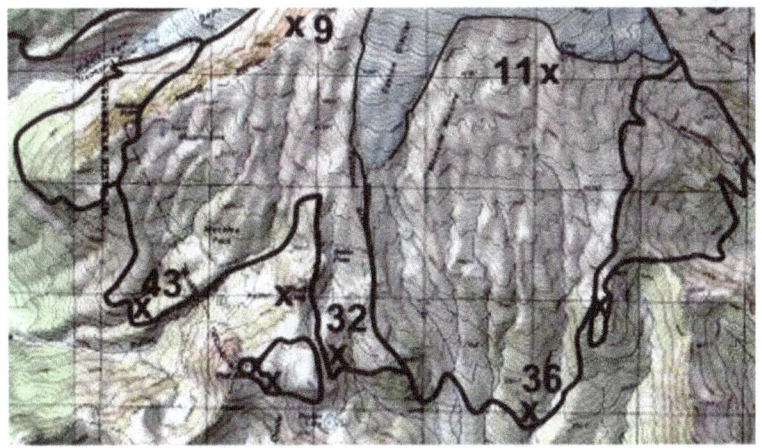

Figure 67. Isotope dates of lava flows on the south flank of Mt. Baker. Numbers are ages in thousands of years. Dark lines are flow boundaries.

Figure 68. North flank of Mt. Baker summit cone (Grant Peak).

MT. BAKER SUMMIT CONE

The main summit of the Mt. Baker volcano is an ice–filled crater about 1,300 feet in diameter standing 10,781 feet above sea level (Figs. 69-75). Radio echo sounding indicates that the ice in the summit crater is at least 300 feet thick. The glacier in the crater flows northward through a gap in the crater wall into the upper Roosevelt glacier. The highest point at the summit, Grant Peak, lies at the east end of the crater where outward–dipping fragmental ejecta interbedded with brecciated lavas are exposed in the sheer east face of the summit crater. Just below the summit of the mountain, coarse breccia makes a cliff known as the Roman Wall.

Figure 69. Mt. Baker volcanic cone. (Photo by Lee Mann)

Figure 70. Looking over Black Buttes (left) to Mt Baker summit cone (left center), Sherman Crater, and Sherman Peak (right).

Figure 71. North side of the Mt. Baker summit cone.

Figure 72. North face of summit crater at the head of the Roosevelt glacier.

Figure 73. Mt. Baker summit cone (top left), steaming Sherman Crater, and Sherman Peak (right). Rocky crags on the left are the Black Butes.

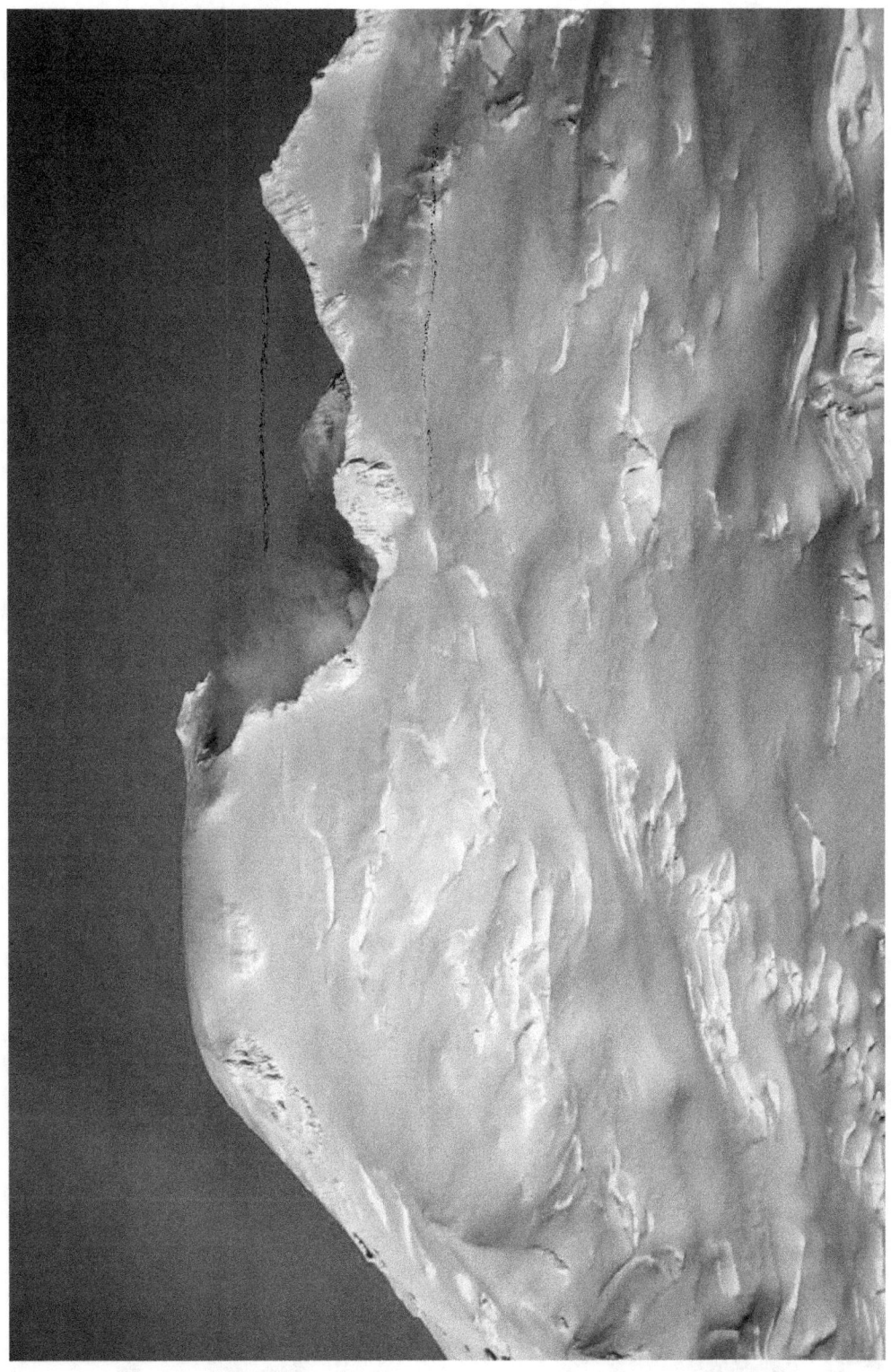

Figure 74. Mt. Baker summit cone (Grant Peak), Sherman Crater, and Sherman Peak.

Figure 75. Mt. Baker summit cone (Grant Peak), steaming Sherman crater, and Sherman Peak.

The glacier that occupies the crater in the summit cone rests on a thick layer of rarely exposed, gray sediments, which rest on older lava flows and fragmental deposits (Fig. 76).

The only known exposure of these sediments occurred in 2009. The origin of the gray sediments is unknown—they could be either volcanic ejecta or glacial deposits.

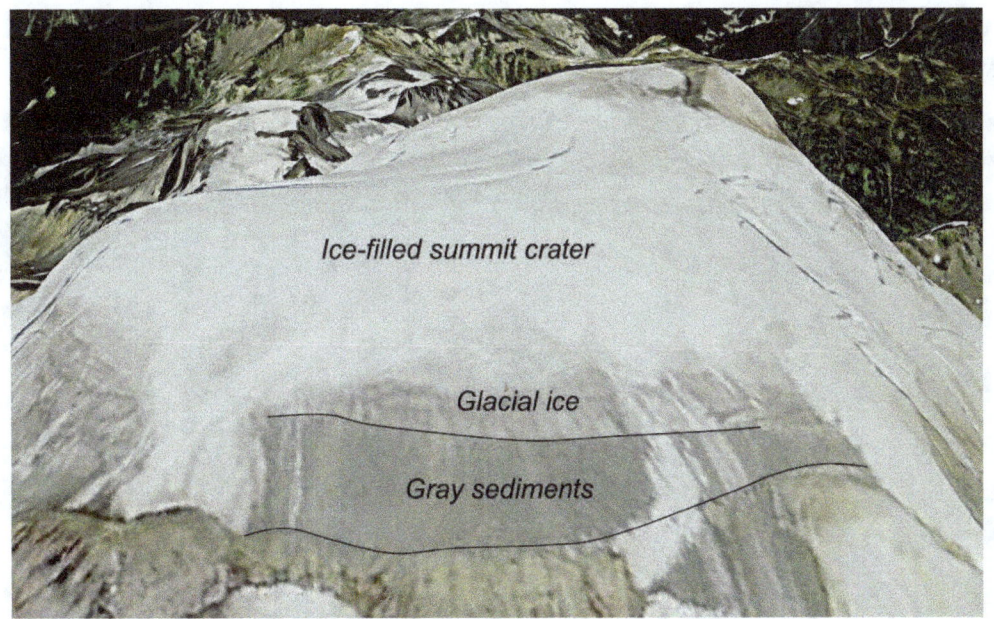

Figure 76. Gray sediments between the summit glacier and underlying lava flows and breccia.

Sherman crater

Sherman crater lies just south of the main summit crater (Figs. 77–80). It is about 2000 feet in diameter and makes a deep circular depression about 1,500 feet below the main summit. The crater is younger than the main summit crater and eruptions from it seem to have been entirely ash and pumice that have mantled the surrounding area. The glacier that occupies the crater floor flows out of the crater through a gap in the east wall of the crater and forms the head of the Boulder glacier.

Sherman crater has long been the site of steam eruptions from vents on the crater floor and around the western and northern margins (see the section on steam eruptions below).

Figure 77. Sherman crater and Sherman Peak. (Photo by Lee Mann)

Figure 78. Sherman crater and summit crater (right). View from the east.

Figure 79. Steam erupting from Sherman Crater, February 1, 1979

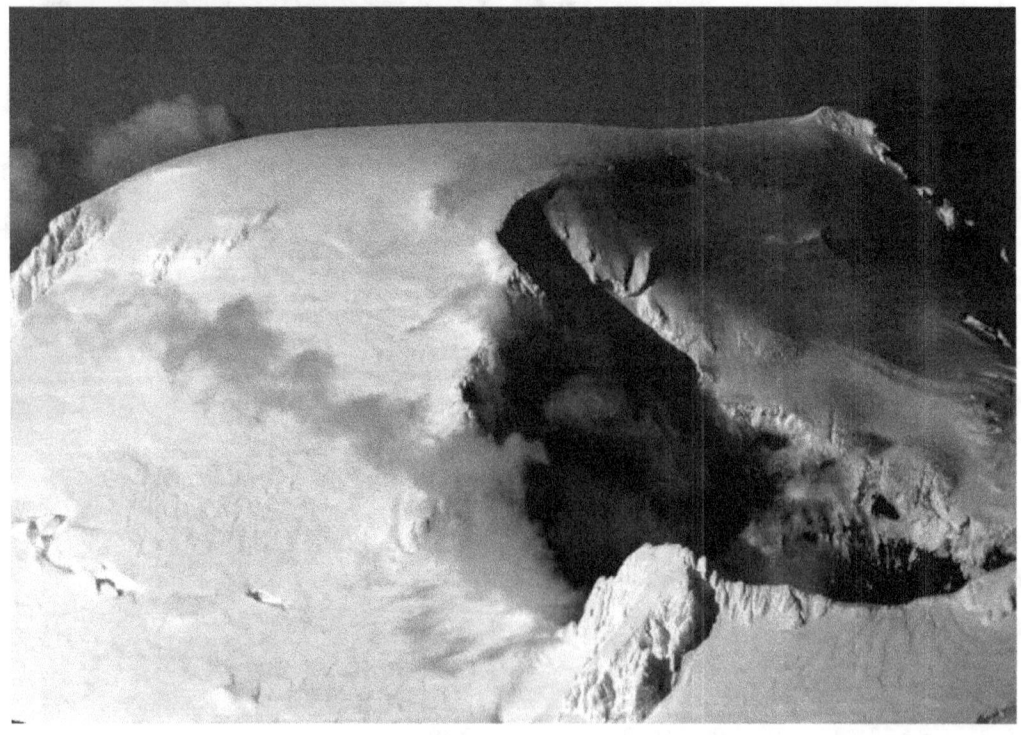

Figure 80. Steam from the west rim of Sherman Crater.

SULPHUR CREEK LAVA FLOWS AND SHREIBERS MEADOW CINDER CONE

Multiple basalt lava flows erupted in the valley of Sulphur Creek below the Easton glacier 8,500 ^{14}C years ago and flowed more than seven miles down Sulphur Creek to Baker Lake (Figs. 81–86). These basalt flows probably came to the surface up the Sulphur Creek fault, which extends the length of Sulphur Creek valley and crosses the divide into the upper Nooksack Middle Fork valley. At least three periods of eruption are represented by lava flows in the Baker River valley, separated by glacial deposits.

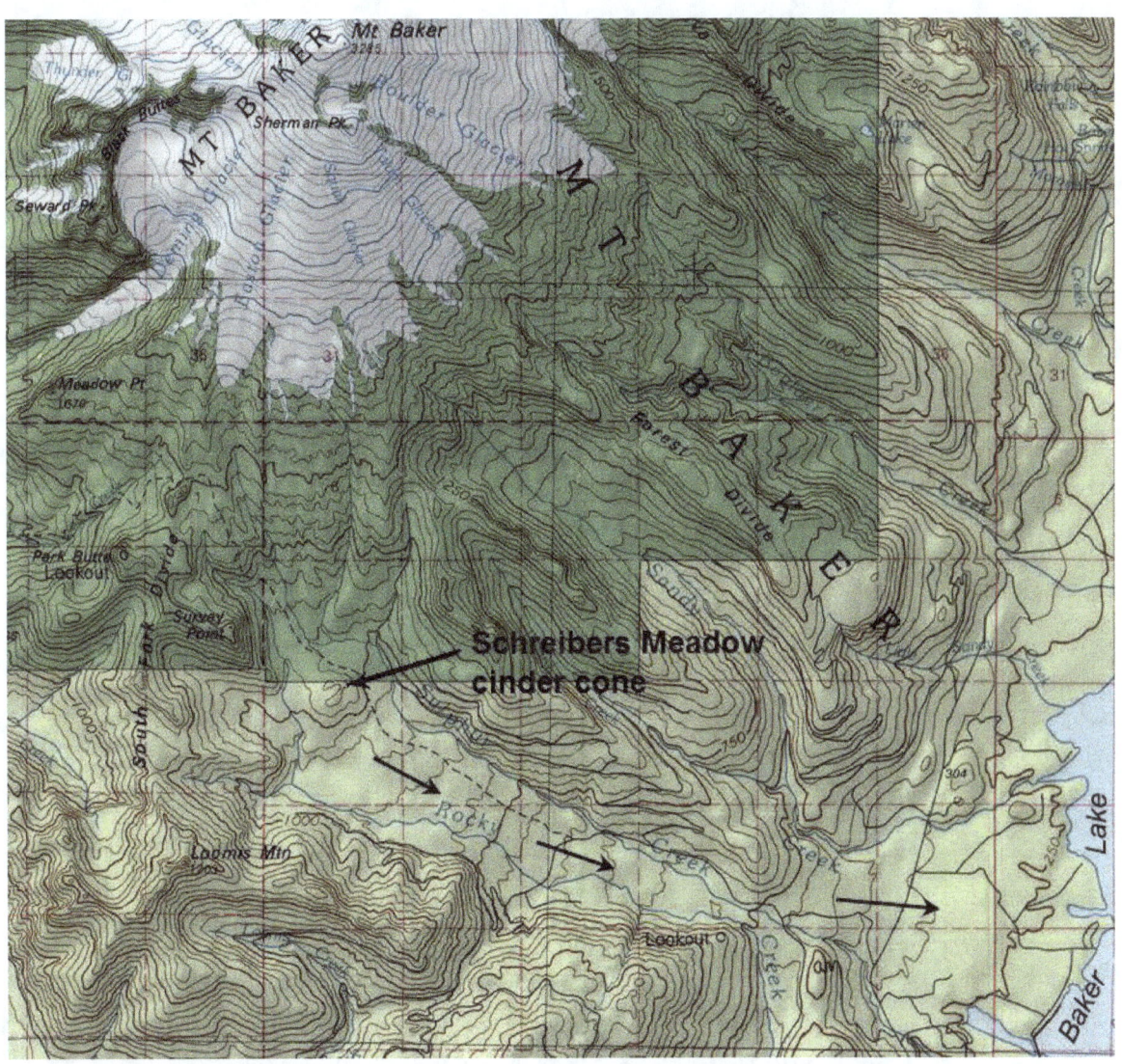

Figure 81A. (Left) Lava flows from Sulphur Creek into Baker Lake.

Figure 18B. Baker Lake, Mt. Baker. (Photo by Lee Mann)

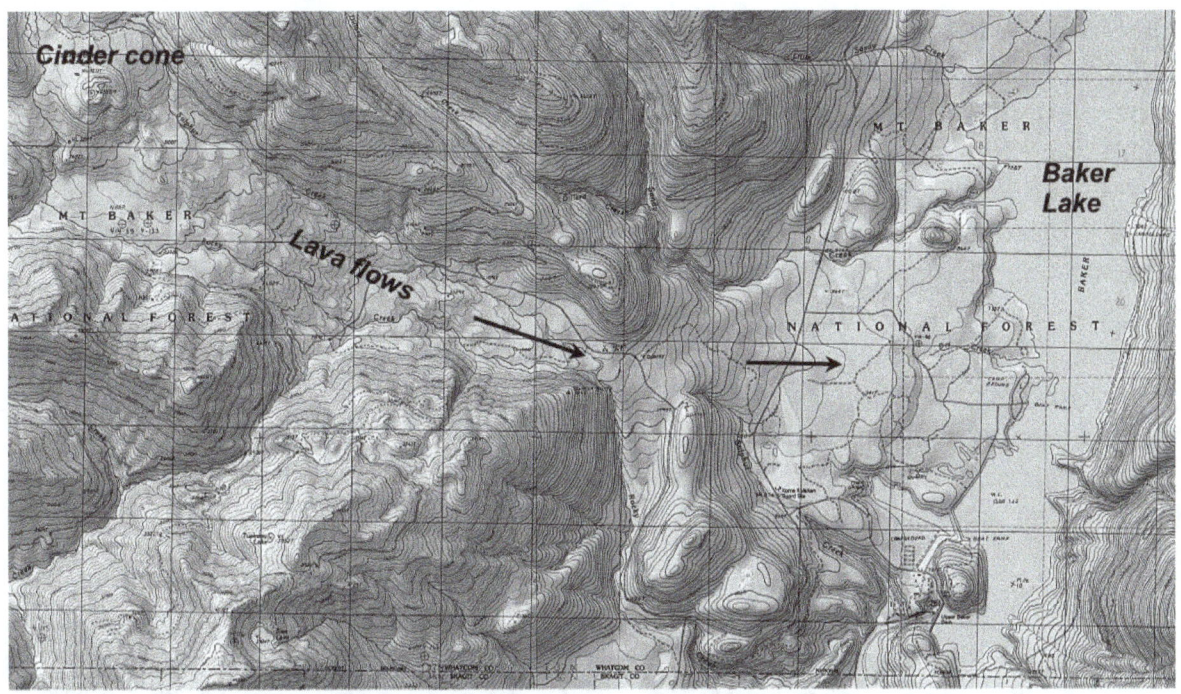

Figure 82. Lava flows in Sulphur Creek valley.

Figure 83. Lava flows down the Sulphur Creek valley to Baker Lake.

The oldest lava flow is buried by about 50 feet of glacial till, which is in turn overlain by 250 feet of younger lava (Lava no.2) (Figs. 84, 85). Lava no.2 makes up most of the flat, fan–shaped area west of Baker Lake (Fig. 86). As the lava flow advanced, it picked up blocks of glacial clay and plastered them against the valley sides. In some places, the lava attained thicknesses up 350 feet. The bottom portion of the flow is dense and highly fractured, whereas the upper part is composed of blocks and clinkers formed as the chilled lava surface was broken up by continued movement of the still–molten lava beneath. The lava flow dammed the ancestral Baker River and displaced it to the east where it cut a new channel, as did Sulphur Creek.

Another, younger lava flow (Lava no.3) erupted from the vicinity of the Schreibers Meadow cinder cone, filled the valley of Sulphur Creek, and partially covered the older lava at what is now Baker Lake. This flow destroyed Sulphur Creek and when drainage from the valley sides re–established the creek along the sides of the valley, two streams, the new Sulphur Creek and Rocky Creek, occupied the valley.

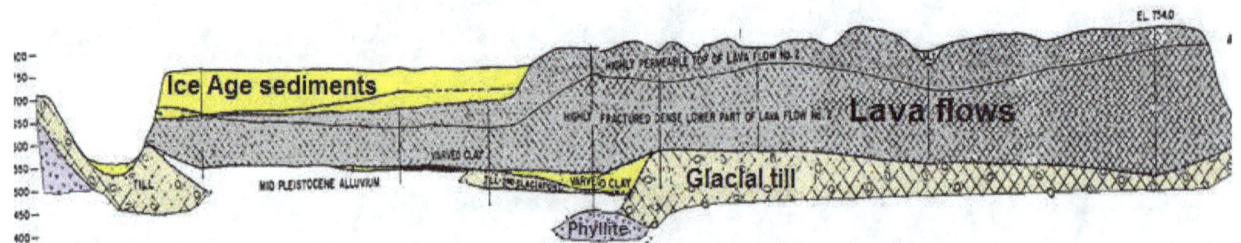

Figure 84. SW to NE geologic cross–section in the Baker River valley between Sulphur Creek and the ancestral Baker River.

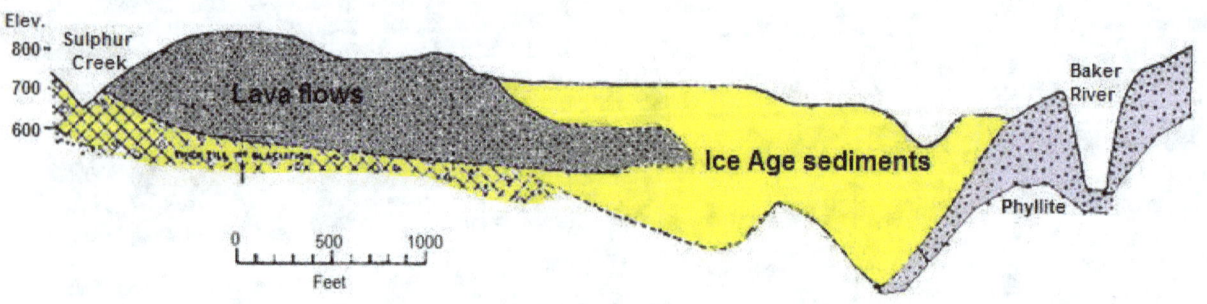

Figure 85. E–W geologic cross–section in the Baker River valley between Sulphur Creek and the ancestral Baker River.

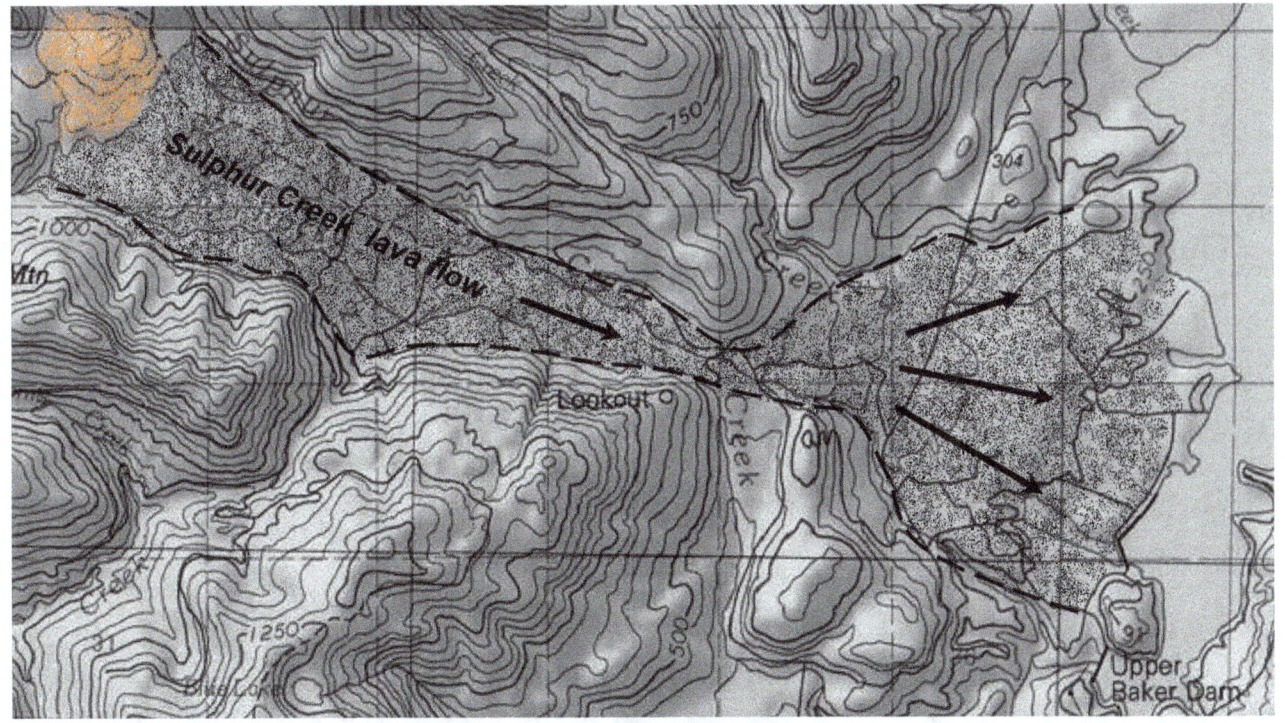

Figure 86. Sulphur Creek lava flow from Schreibers Meadow cinder cone to Baker Lake.

The volcanic eruption at Schreibers Meadow began with ejection of a distinctive, coarse–grained, reddish–orange scoria (lava with many holes from escaping gas bubbles) near the vent (Figs. 88, 90, 91) and deposition of coarse air–fall ash on the southwestern flank of Mt. Baker along a northeasterly fallout plume. The scoria is up to six feet thick near the vent, but a fair amount of it has accumulated near the base of slopes by down–slope movement from the slopes above. The scoria thins to the northeast where meadows that have not been overrun by glaciers show a distinctive orange color (Figs. 90, 91). The scoria thins to three feet about a mile from the vent and is only about 2–4 inches thick five miles northeast of the cone. Pieces of pumice up to an inch in diameter occur near the cinder cone, but decrease to about a quarter of an inch near the end of the fallout plume. At its type locality in an alpine meadow just south of the Squak Glacier, the scoria contains abundant charcoal, which was radiocarbon dated at 8,420 ± 70 radiocarbon years old. The scoria is overlain by the Sulphur Creek lava flow and a volcanic mudflow radiocarbon dated at 8,500 radiocarbon years (Fig. 87).

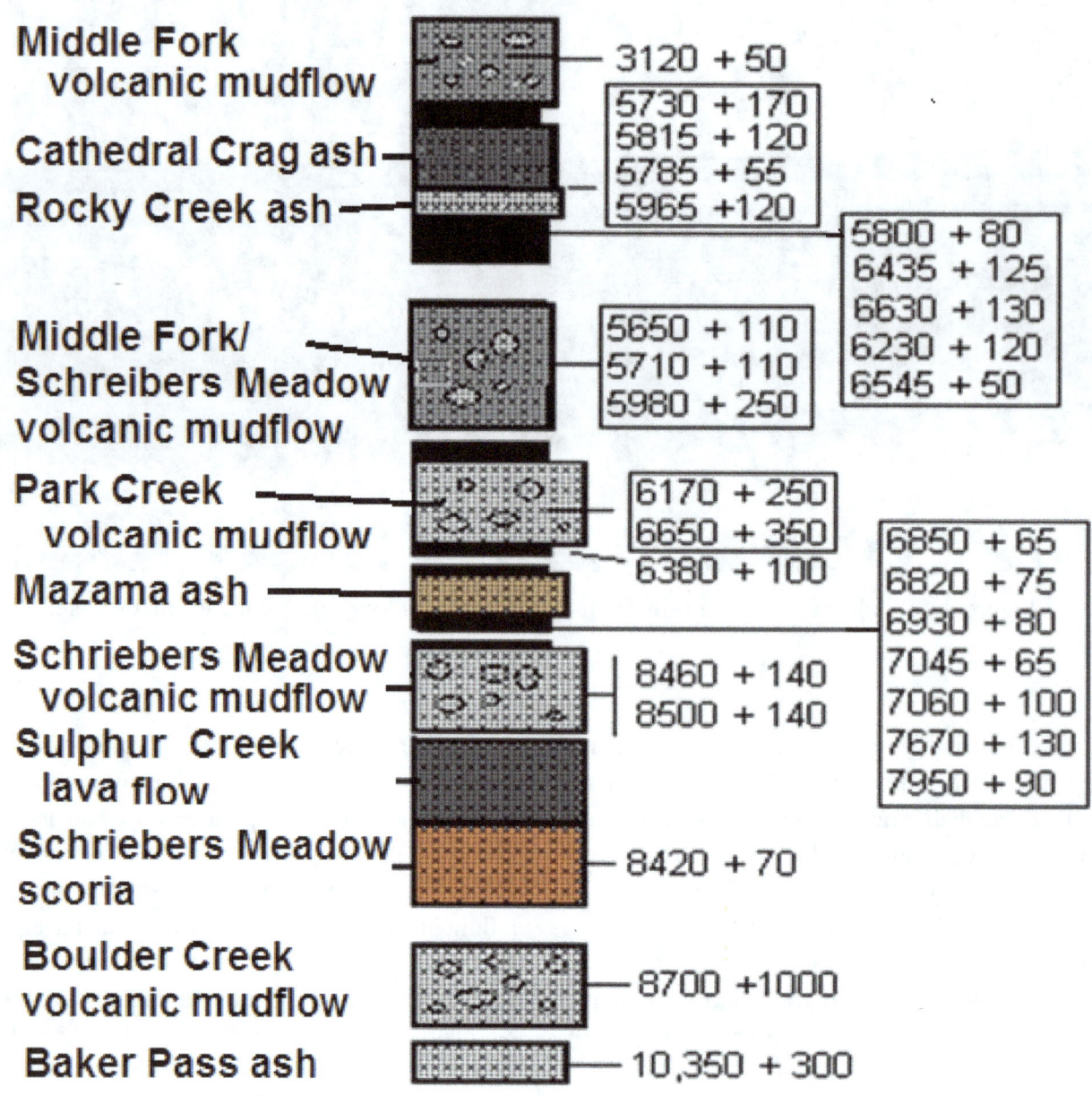

Figure 87. Chronology of volcanic ash eruptions and mudflows for the past 10,000 years.

Figure 88. Schreibers Meadow scoria in a roadcut near the meadow.

Figure 89. Location of sites on the south flank of Mt. Baker. Numbers are geologic sites: type localities of the Rocky Creek and Cathedral Crag ashes, 12, 14; type locality of Schreibers Meadow scoria, 30.

Figure 90. Schreibers Meadow scoria below the Squak glacier on the SW flank of Mt. Baker.

Figure 91. Schreibers Meadow scoria NE of Schreibers Meadow.

Figure 92 Schreibers Meadow cinder cone.

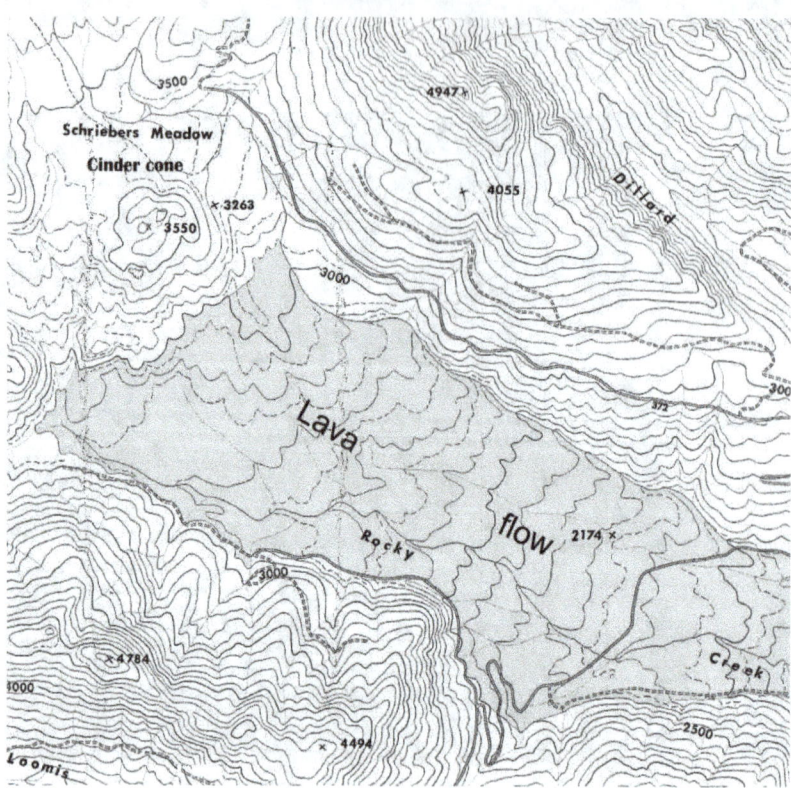

Figure 93. Schreibers Meadow cinder cone and Sulphur Creek lava flow.

VOLCANIC ASH ERUPTIONS

Five volcanic ashes have been recognized in meadow exposures around Mt. Baker—Schreibers Meadow scoria, Mazama ash, Rocky Creek ash, Cathedral Crag ash, and the 1843 Mt. Baker ash. The oldest of three ash layers in the Boulder Creek valley on the south flank of Mt. Baker lies beneath two lava flows, two other ash layers, and other volcanic deposits. Its thickness and particle size suggest that it originated from Mt. Baker (Fig. 94). A small wood fragment beneath the ash was dated at 8,700 ± 1,000 years, but the date is tentative because of the limited amount of sample material available. A younger ash, separated from the one described above by a lava flow, is less than an inch thick and consists of very fine particles, suggesting a distant source. A still younger fine-grained ash layer in Boulder Valley, overlain by a lava flow, is ~3 inches thick.

Figure 94A. Volcanic and glacial deposits in Bouolder valley.

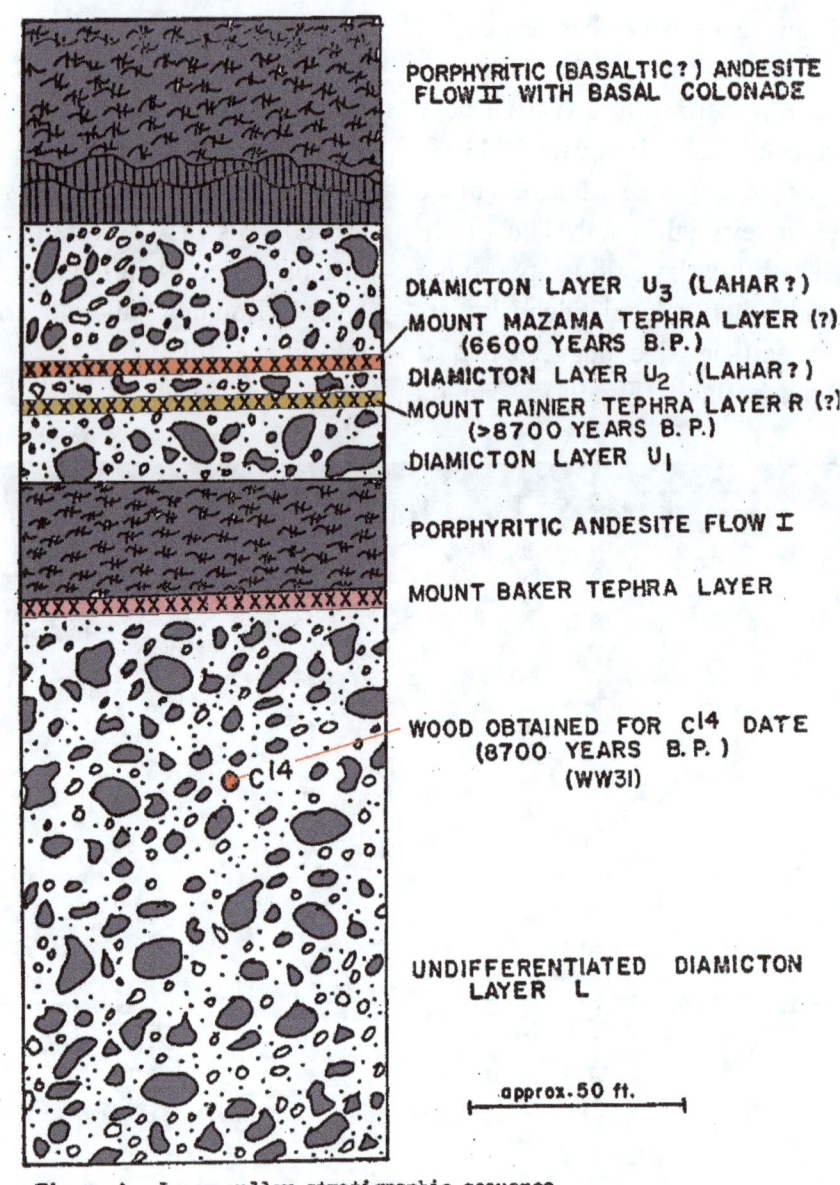

Figure 94. Stratigraphic section on the east side of Boulder Creek. (Burke, 1972)

Figure 95. Mazama ash (yellow layer above ice axe), Rocky Creek ash (white layer), and Cathedral Crag ash (black layer under heather) at Heather Meadows.

Mazama Ash

The cataclysmic eruption of Mt. Mazama in southern Oregon covered the entire Pacific Northwest with several inches of volcanic ash and subsequent collapse of the volcano created Crater Lake. Mazama ash is exposed along trails and creek banks in almost every meadow around Mt. Baker (Figs. 95–98). It typically consists of fine–grained, yellowish ash, 1-3 inches thick, composed mostly of volcanic glass with abundant pumice. The ash is chemically distinct from Mt. Baker ashes and can be readily recognized both in the field and in lab. Radiocarbon dating of peat above and below the ash at seven localities establishes it's age at 6,850 radiocarbon years.

Figure 96. Mazama ash, Rocky Creek ash, and Cathedral Crag ash at Heather Meadows.

Rocky Creek Ash

A light–gray, sandy ash occurs above Mazama ash at Heather Meadows, Artist Point, Ptarmigan Ridge, Chain Lakes, and many other meadows around Mt. Baker (Figs. 95–98). It was first recognized and dated along the Chain Lakes trail in 1975 and later named the Rocky Creek ash at its type locality along the Schreibers Meadow trails on the SW flank of Mt. Baker (Fig. 97) where it is underlain by peat and Mazama ash and overlain by Cathedral Crag black sandy ash.

Rocky Creek ash consists mostly of well–sorted, sand/silt–size, rock particles with minor volcanic glass. The ash is typically a few inches thick, but in some places reaches thicknesses up to 12 inches. Microprobe analyses of the chemical composition of glass in the ash indicate that it was erupted from Mt. Baker. The composition is definitely different from volcanic glass in Mazama ash but quite similar to the overlying Cathedral Crag ash. The age of the Rocky Creek ash is tightly constrained at close to 5,800 radiocarbon years by dates of 5,785 ± 55, and 5,800 ± 80 radiocarbon years below the ash and by dates of 5,730 ± 170, 5,815 ± 120, and 5,965 ± 120 radiocarbon years above the ash.

Figure 97. Rocky Creek and Cathedral Crag ashes at their type locality along the Schreibers Meadow trail on the south flank of Mt. Baker.

Cathedral Crag Ash

The Cathedral Crag ash was defined and dated at its type locality along the Schreibers Meadow trail on the SW flank of Mt. Baker (Fig. 97). It consists of massive, black, sandy ash and is the thickest and most widely distributed ash erupted from Mt. Baker. It is up to 30 inches thick at Heather Meadows and makes a prominent peninsula extending into south Picture Lake. The ash consists of well sorted, crystal–rich, coarse sand-sized particles. Microprobe composition of the glass resembles other Mt. Baker ashes, but is distinctly different from Mazama ash.

Organic material beneath Cathedral Crag ash has been dated at 5,785 ± 55, and 5,800 ± 80 radiocarbon years. A date of 5,780 years was obtained from plant fragments under the ash in upper Swift Creek between Artist Point and Mt. Shuksan. These dates are very close to the 5,800 radiocarbon year age of the underlying Rocky Creek ash.

Because Cathedral Crag ash typically lies directly on Rocky Creek ash with no intervening peat or sediment, the Rocky Creek ash apparently resulted from a steam eruption carrying mostly rock particles ripped from the sides of the vent, closely followed by eruption of the Cathedral Crag ash containing volcanic glass from erupting lava that cooled in the air.

Figure 98. Cathedral Crag, Rocky Creek, and Mazama ashes above Schreibers Meadow.

HISTORIC ERUPTIONS OF MT. BAKER

When Captain Vancouver sailed into Puget Sound in April, 1792, he observed a high volcano that was emitting "smoke" (actually steam), In June 1792, the Spanish expedition of Dionisio Alcalá Galiano and Cayetano Valdés, while anchored in Bellingham Bay, reported: *"During the night, we constantly saw light to the south and east of the mountain of Carmelo [Mt. Baker] and even at times some bursts of flame, signs which left no doubt that there are volcanoes with strong eruptions in those mountains."* In 1843, explorers reported widespread ash that fell "like a snowfall" and that the forest was "on fire for miles around." Indians reported that nearby rivers were clogged with ash and many salmon perished. Gray silt-sized ash and rock fragments up to five inches in diameter are scattered on the slopes between the Deming and Easton glaciers. Most of the larger rock fragments have been chemically altered and are coated with sulphur. It was most likely erupted from Sherman Crater by a steam eruption of Mt. Baker in 1843 that ripped fragments from the sides of the vent. No lava flows were associated with the eruption.

In 1854, George Davidson of the Coast and Geodetic Survey noted *"the summit of this mountain obscured by vast rolling masses of dense smoke, which in a few minutes reached an estimated height of two thousand feet above the summit, and soon enveloped it entirely."* In 1858, steam activity was noted by several people and on November 26, 1860, passengers traveling by steamer from New Westminster to Victoria reported that Mount Baker was *"puffing out large volumes of smoke, which upon breaking, rolled down the snow-covered sides of the mountain, forming a pleasing effect of light and shade."*

Steam eruptions

During the first ascent of Mt. Baker by the Coleman party in 1868, climbers reported emission of steam from several fumaroles in Sherman Crater. On July 8, 1891, a climbing party reported, *"The opening was fifty by seventy-five feet in circular shape and puffs of smoke issuing from the interior would vanish in the light air at a short distance."* On September 9, 1891, another climbing party observed that the crater " . . . *"is filled with snow, except in the center, where there is a circular opening from which steam and sulphurous vapors are constantly escaping. In fact we noticed the presence of the sulphurous vapors 2000 feet below the summit."* John A. Lee described 'smoke' rolling up from the crater in great clouds and a large round vent in the snow from which steam and sulphur fumes were issuing.

In 1906, C. F. Easton reported that *"vapors of sulphur are continually emitted and steam jets issue from myriads of vents with great violence."* Since then, the crater has been seen by numerous climbing parties and observed from airplanes, but published reports of activity are scarce. Air photographs taken in 1947, 1950, and 1955 show some steam activity or glacial melting, and mild steam activity in the early 1960s.

On March 10, 1975, a tall column of steam was reported rising some 1,500 feet above the floor of Sherman Crater. Weather conditions prevented observation from the air, but cleared the following day. Dave Rahm, a geologist and aerobatic pilot, and I flew over the crater March 11 and continued aerial photography until 1982. We would take the doors off the plane to get clearer photos, and Dave would dive into the crater and flip the wings to vertical to get an unobstructed view of the steam vents at close range.

On March 11, 1975, a powerful steam jet was ejecting steam to heights of 300 to 800 feet above the crater floor of Sherman Crater. A circular vent about 100 feet in diameter in the glacier near the north wall of the crater was discharging a column of steam and the glacier on the floor of the crater was covered with black ash (Figs. 99,100).

Figure 99. Ash–covered floor of Sherman crater, March 11, 1975. Steam emanating from the north–west (top) and east rim (lower). The ash was erupted during a more vigorous eruption on March 10.

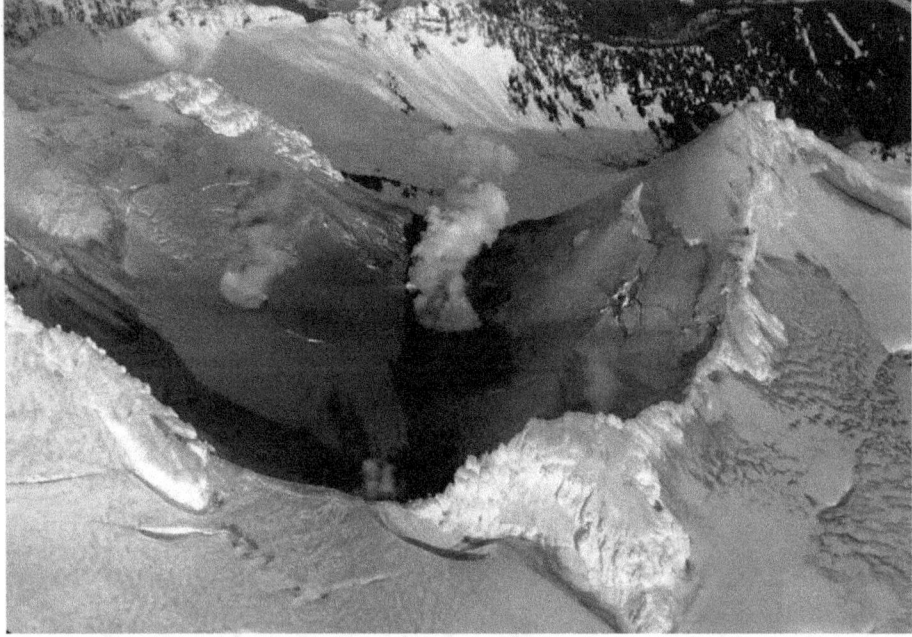

Figure 100. Ash–covered glacier on the floor of Sherman crater. Principal areas of steam activity were the north vent against the north wall of the crater (left vent), the east vent at the breach in the crater (right vent), and a series of vents on the west rim. Another vent later opened in mid–glacier on the crater floor.

In early April, 1975, the glacier at the north wall of Sherman Crater developed large crevasses as ice pulled away from the rest of the glacier and fell into the vent below. A large elongate pit about 300 feet deep developed in the ice at the base of the north wall, causing the glacier above to collapse down into the vent area (Figs. 101–103). By April 1, 1975, a depression about 200 feet in diameter bounded by a circular fracture appeared in the glacier near the center of the crater floor and steam vents were continuously active at the east rim of the crater (Figs. 101–103).

Figure 101. Steam vents on the floor of Sherman Crater, April 1, 1975. The circular crevasse in the glacier (upper right) is formed by the collapse of ice from melting above a steam vent.

Figure 102. Steam vents on the floor of Sherman Crater, April 1, 1975. Glacial ice is collapsing (right) above a steam vent.

By May 7, 1975, the roof of the central depression had completely collapsed, making a ~150–foot deep hole in the glacier, revealing a lake on the crater floor and a steam vent along the side. (Fig. 103). Throughout following months, parts of the lake could occasionally be observed boiling. Flights over the crater every couple of weeks showed that the central lake, east rim, and west rim vents remained about the same, but the north vent deepened and expanded (Figs. 105–106).

Figure 103. Lake in the circular depression melted by steam at the central vent shown in figures 100 and 101, May 7, 1975.

An unusually large stream eruption, clearly visible from Bellingham (Fig. 104), occurred June 29, 1975. Flying over the mountain the next day revealed that the north vent had deepened and expanded and the glacier on the crater floor had become heavily crevassed as ice flowed toward the melting area around the vent (Figs 105–106). The glacier on the north wall of the crater was also heavily crevassed as the ice collapsed down into the vent area (Fig. 105)

Figure 104. Steam rising from Sherman Crater at sunset.

Figure 105. Steam rising from the north vent against the north wall of the crater June 30, 1975. The glacier on the crater floor has broken into large blocks along crevasses as the ice collapsed toward the vent. The glacier on the north wall above the vent is also heavily crevassed as the ice collapsed into the vent.

Figure 106. Looking down the north vent June 30, 1975. Ice above the vent is collapsing into it.

A week later, July 9, the glacier on the crater floor was covered with a thick blanket of black volcanic ash (Figs. 107–108) that erupted between June 30 and July 9.

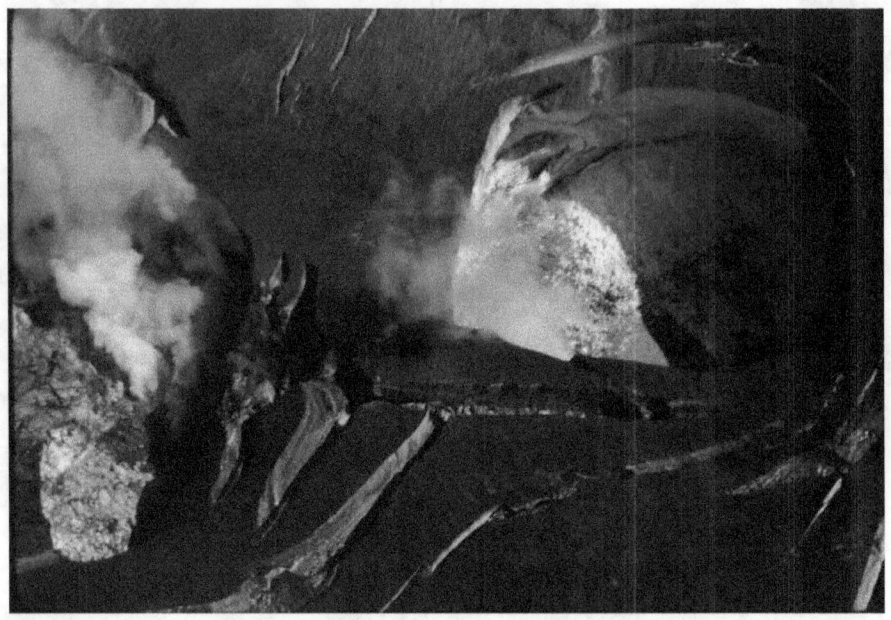

Figure 107. Blanket of black volcanic ash on the crater floor glacier July 9, 1975. The deep pit on the right is the central vent formed by the collapse of ice from melting. The open crevasses in the glacier have formed as ice flowed toward the open space. The steaming vent on the left is the east rim vent.

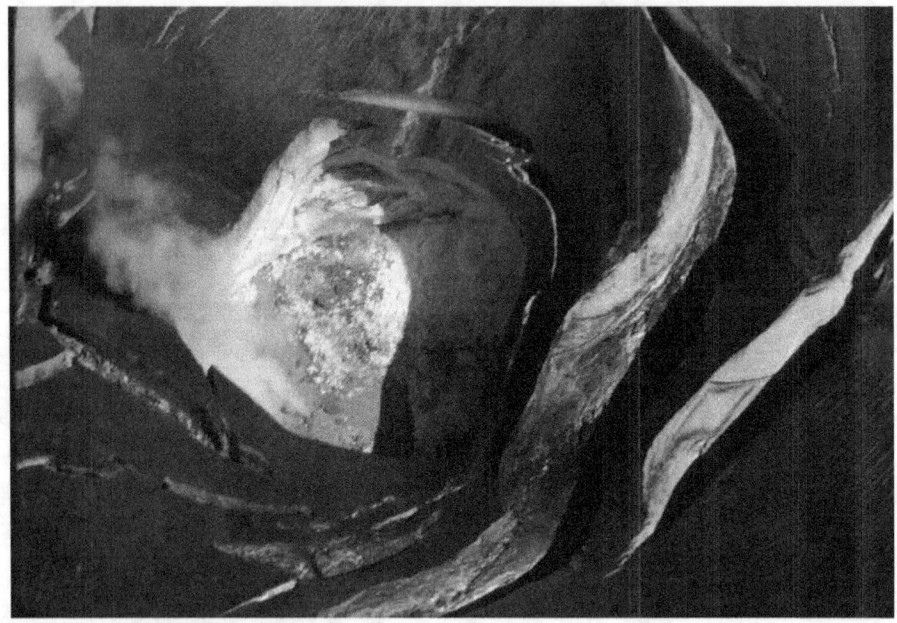

Figure 108. Close-up of steam rising from the central vent July 9, 1975. Large crevasses open up as the ice flows toward the melting ice in the vent area. The surface of the glacier is covered with black volcanic ash from the eruption between June 30 and July 9, 1975.

The hole in the crater floor glacier continued to enlarge (Fig. 109) as ice flowed into the open space, caved off, and melted.

Figure 109. Enlargement of the central vent area, September 21, 1976 until it nearly merges with the east rim vent. The yellow color near the vent on the right side is sulphur at the east rim vent.

Another eruption of black ash occurred sometime between October, 1976 and January, 1977, covering the glacier once again with black ash (Fig. 110).

Figure 110. Ash covered glacier on the crater floor January, 1977. Steam activity at the east rim vent has continued to enlarge the ice–free area and enlargement of vent area at the north vent has caused increasing collapse of the glacier above it. Bare areas in the northwest part of the crater (left side) have appeared as a result of steam activity there.

On November 21, 1978, an unusually large steam eruption occurred along the west rim (Figs. 111–116), filling Sherman crater with steam.

Figure 111. Large steam eruption along the west rim of Sherman crater November 21, 1978.

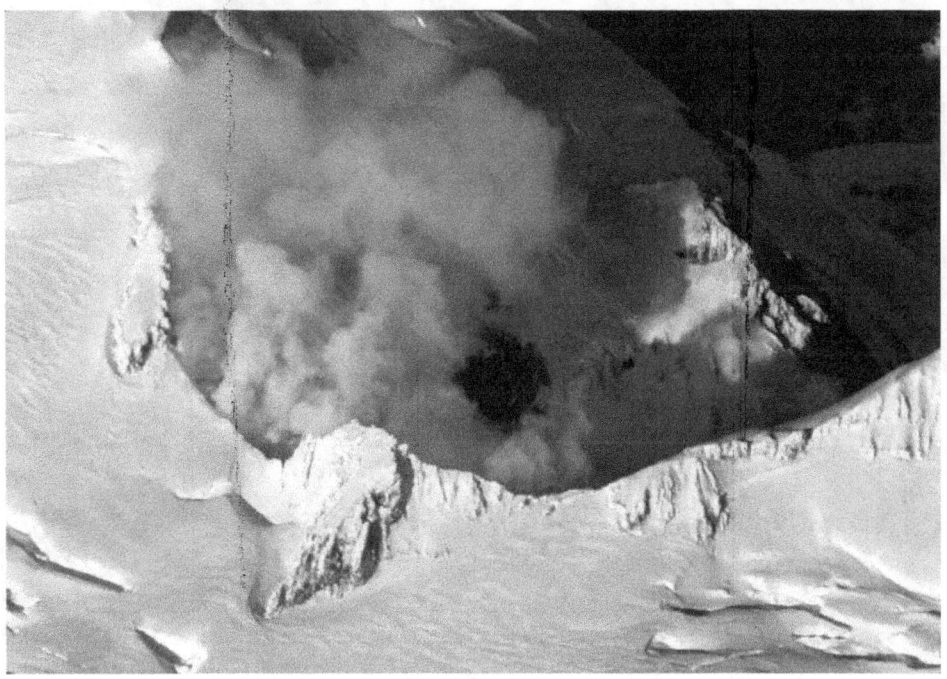

Figure 112. Large steam eruption Nov 21, 1978.

Figure 113. Steam rising the west rim of Sherman crater November 21, 1978.

Large steam eruptions from the west rim and northwest crater wall occurred January 29 and February 28, 1979 (Figs. 111-115).

Figure 114. Steam eruption from the west rim January 29, 1979.

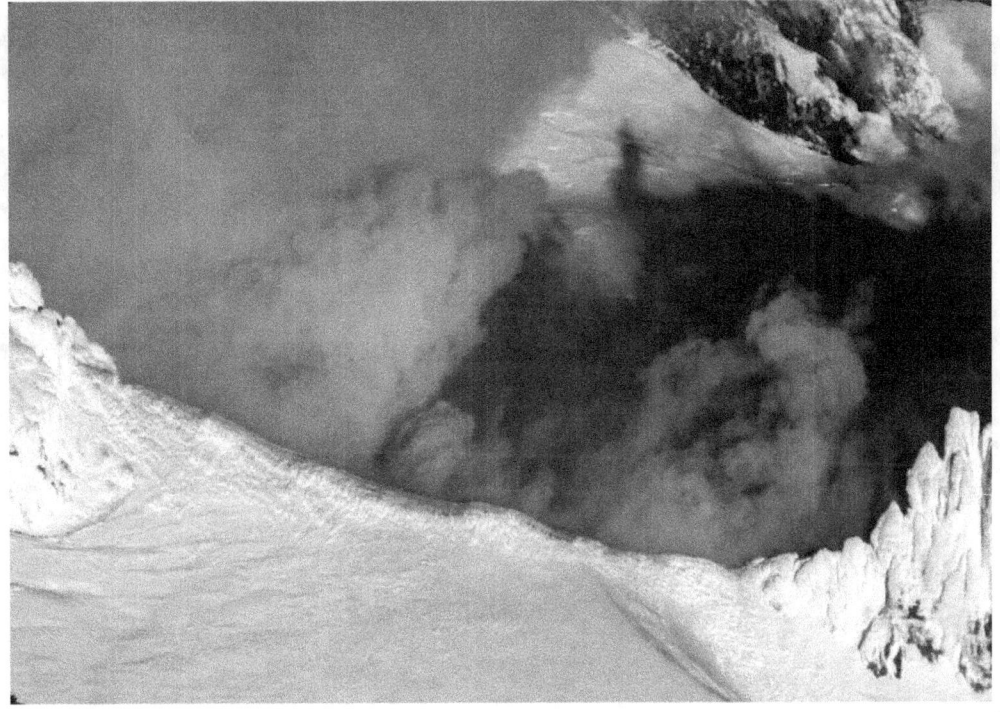

Figure 115. Large steam eruption from the west rim February 28, 1979.

A vigorous steam eruption from the west rim and northwest wall of the crater occurred Jan 28, 1980 (Fig. 116). Since the early 1980s, steam activity has subsided, although it still continues, and the hole in the glacier at the central vent area has closed up with no further sign of steam there.

The Dorr vent area at an elevation of 7800 feet on the north flank of Mt. Baker consists of many small vents that emit sulphur-charged vapor along a ridge of altered rock.

Figure 116. Large steam eruption from Mt. Baker viewed from Bellingham.

A large mass of snow and rock debris that broke away from the steep cliffs at the east end of Sherman Peak (Fig. 117) and flowed down the Boulder glacier (Fig. 118-120) was apparent on July 26, 1979. A similar avalanche occurred in 2006.

Figure 117. Scar from an avalanche of snow and rock from the east face of Sherman Peak, July 26, 1979.

Figure 118. Path of rock debris from snow and rock slide on the east side of Sherman Peak. (1979)

Figure 119. Rock debris from debris avalanche off the east face of Sherman Peak. July 26, 1979.

Figure 120. Rock debris on the Boulder glacier from a snow and rock avalanche off the east face of Sherman Peak. 1979.

Volcanic mudflows

Large volumes of ash and fragmental material accumulate on the flanks of volcanoes. As steam and hot water associated with volcanic heat migrate through rock, they chemically alter and decompose it. Both the ash and chemically altered rock are unstable on steep slopes and are prone to slope failures, especially if they are saturated with water from melting of glacial ice. Failures of such unstable slopes produce volcanic mudflows (lahars) of water–saturated mud, ash, and rock fragments swept up by the mudflows that can flow many miles downvalley. An example of a large–volcanic mudflow is the Osceola mudflow that flowed down Mt. Rainier into the Puget Lowland about 5,000 years ago and deposited about 100 feet of mud and debris in south Seattle. Similar, smaller volcanic mudflows have occurred at Mt. Baker over the past several thousand years.

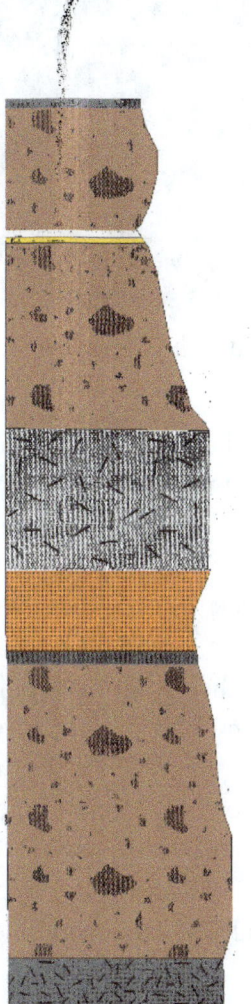

Cathedral Crag ash
Volcanic mudflow ~5,800 ^{14}C yrs BP

Rocky Cr. ash ~5,800 ^{14}C yrs BP
Mazama ash 6,850 ^{14}C yrs BP

Volcanic mudflow ~8,500 ^{14}C yrs BP

Sulphur Creek lava flow

Schreibers Meadow scoria
~8,800 ^{14}C yrs BP
Baker Pass ash

Volcanic mudflow

Andesite lava flow

Figure 121. Volcanic mudflows (lahars) and lava flows at Schreibers Meadow.

Figure 122. Volcanic mudflows and ash at Schreibers Meadow.

The Boulder Creek valley on the south flank of Mt. Baker is filled with volcanic mudflows, which have built a large delta into the Baker River valley, now largely covered by Baker Lake. Wood from one of the oldest of these mudflows has been radiocarbon dated at 8,700 ± 1000 radiocarbon years. Volcanic mudflow deposits containing charred wood at Schreibers Meadow, radiocarbon dated at 8,460 ± 140 and 8,500 ± 140 radiocarbon years, are overlain by Mazama ash and Rocky Creek ash and another lahar radiocarbon dated at 5,800 radiocarbon years (about 6,600 calendar years) (Fig. 122).

The 5,800 year old lahar originated on the flanks of Mt. Baker and flowed at least 20 miles down the Middle Fork of the Nooksack River valley Figs. 123, 124). The mudflow was so thick it flowed far up into Clearwater Creek, a tributary of the Middle Fork, reaching 130 feet above the level of the Middle Fork. Many large logs in the mudflow in the Middle Fork near the mouth of Clearwater Creek and at the confluence of the Nooksack Middle and North Forks in the area near Welcome (Fig. 124) are oriented parallel to one another pointing up the Middle Fork valley, indicating the direction of flow. Much of the wood is charred, suggesting that the lahar may have been hot. Many of the rocks in the mudflow are rimmed with sulphur or are hydrothermally altered, indicating its volcanic origin. Wood in the mudflow has been dated at 5,710 ± 110 and 5,650 ± 110 radiocarbon years (Fig. 122). The Rocky Creek ash, dated at about 5,800 radiocarbon years, directly underlies the mudflow in places, suggesting that both are related to an eruptive phase of Mt. Baker.

This mudflow not only flowed down the Middle Fork valley, but it also spilled down the Sulphur Creek valley on the south flank of the mountain where it forms an irregular surface around the Schreibers Meadow cinder cone and covers the Sulphur Creek lava flow and Schreibers Meadow scoria. Remnants of the mudflow occur on the divide between the Middle Fork and Schreibers Meadow, suggesting that the mudflows at both places belong to the same event.

Volcanic mudflows also flowed down the Park Creek valley. A 15–foot–thick mudflow exposed on the north side of the road at lower Park Creek along the road to Baker Hot Springs about half a mile from the junction with the Baker Lake Road, contains hydrothermally altered rocks up to six feet in diameter. Wood in the mudflow has been dated at $6,650 \pm 350$ radiocarbon years. A mudflow about 20 feet thick, exposed on the east side of the Park Creek fan near the mouth of Swift Creek, contains abundant wood dated at $6,170 \pm 250$ radiocarbon years. A stump buried by a much younger mudflow along the banks of Park Creek has been dated at 530 ± 200 radiocarbon years.

A younger volcanic mudflow is exposed in the floor of the Nooksack Middle Fork channel about 15 miles downvalley from the headwaters. Wood in it has been dated at $3,120 \pm 50$ radiocarbon years. This is the only known occurrence of the mudflow, but because of the distance downvalley from the source, it must have been a substantial flow.

Several volcanic mudflows in the Boulder Creek valley have occurred in the past few hundred years. Two large masses of rock debris have avalanched into the upper part of Rainbow Creek valley, leaving forest trimlines. The older of the two occurred in the mid-1860s and moved at least six miles downvalley. Tree rings from an avalanche–damaged tree indicate that the younger event occurred in 1888.

Volcanic mudflows from Mt. Baker pose a significant hazard along the upper reaches of the Nooksack River. Future lava flows are unlikely to flow far enough downvalley to impact populated areas, but mudflows have extended downvalley as far as Deming. Floods from melting of glacial ice during an eruption could reach the entire length of the Nooksack River. No ash, volcanic mudflows, or other volcanic deposits from Mt. Baker have been found in peat bogs and other sediments as old as 100,000 years anywhere in the lowland of Whatcom County.

Figure 123. Logs dated at 5,800 ^{14}C years year old in a volcanic mudflow in the Middle Fork valley near the mouth of Clearwater Creek.

Figure 124. 5,800–year–old logs in the volcanic mudflow at the confluence of the Middle Fork and North Fork of the Nooksack River.

Figure. 125 Typical massive, poorly sorted, volcanic mudflow in the Middle Fork valley.

GLACIATIONS OF MT. BAKER

Glaciers are masses of ice and granular snow formed by compaction and recrystallization of snow, lying largely or wholly on land and showing evidence of past or present movement. Since ice movement is one prerequisite for a glacier, permanent snowfields that persist through the summer melt season are not considered glaciers because they do not move. Glacial ice begins as snow that survives the warm summer melt season and is gradually transformed into glacial ice by compaction, pressure melting, and refreezing.

Glaciers are fed largely by winter snow, and ice is lost mostly by summer melting or breaking off at the margin. In the upper part of a glacier, accumulation of snow and ice is greater than summer melting. In the lower part of a glacier, melting of ice is greater than accumulation of winter snow, so there is net loss of ice in this part of a glacier. Thus, the lower part of glaciers might be expected to melt away each summer, but they don't because the ice lost to the glacier in the lower zone is replaced by movement of ice from the upper accumulation zone. Thus, the position of the terminus of a glacier is determined by the relative amount of winter accumulation and amount of summer melting. If snow accumulation exceeds melting over a number of years, the amount of glacier ice will increase and the glacier terminus will advance. If summer melting exceeds winter snowfall over time, the glacier terminus will recede.

Thus, glaciers act like nature's thermometers. They fluctuate back and forth, much like mercury in a thermometer. When the climate warms, glaciers recede; when the climate cools, glaciers advance. Although glaciers are also affected by precipitation (snowfall), summer temperatures strongly affect glacier melting and fluctuations of glacier termini. Because glaciers persist for thousands of years and leave a record of their former margins, they are very useful for determining climate changes in the past. Mt. Baker glaciers have been especially useful in reconstructing past climate changes because they have been especially sensitive to climate changes. Glaciers are to climate what mercury is to thermometers.

Mt. Baker has more glaciers than any other mountain in Washington except Mt. Rainier. It is one of the snowiest places on Earth, holding the world record for snowfall in a single season, 1,140 inches (95 feet) in 1999.

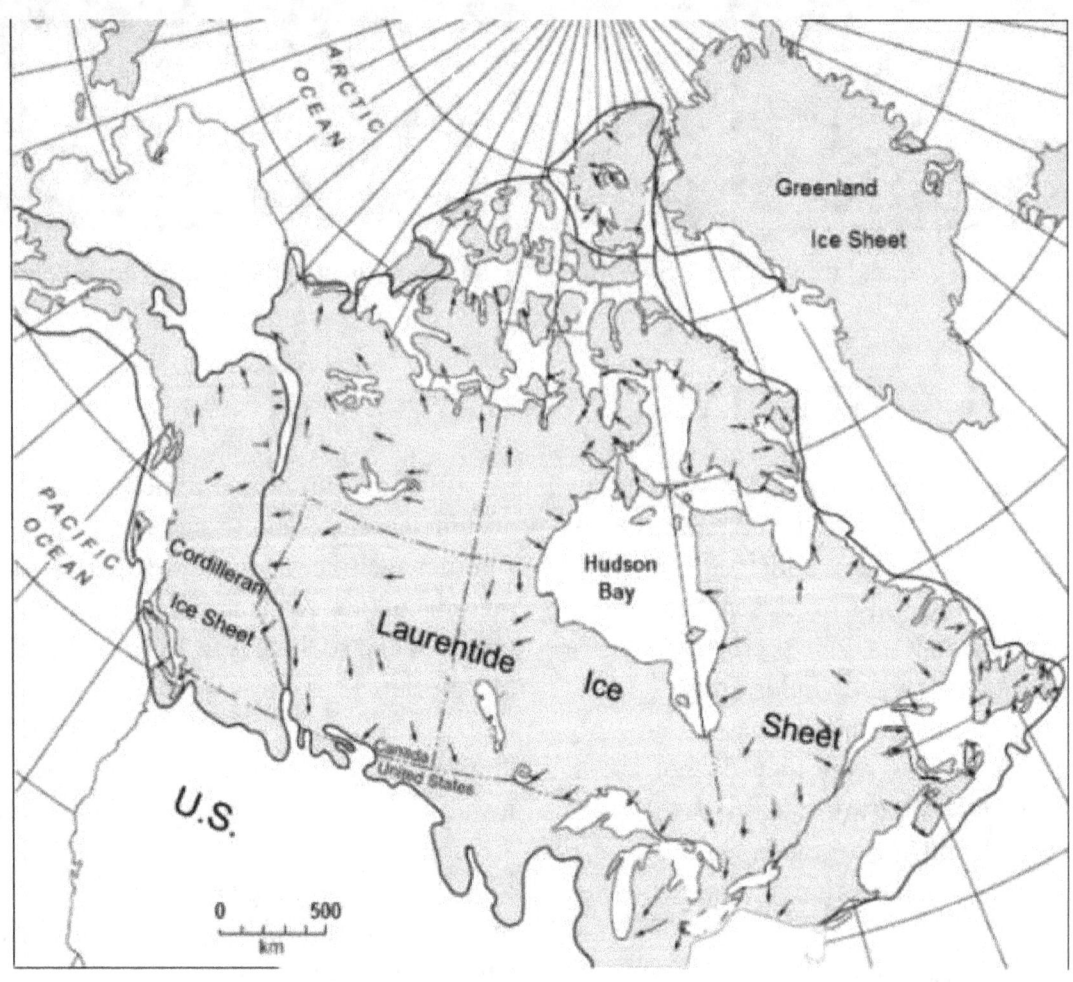

Figure 126A. Huge continental glaciers of the last Ice Age in North America

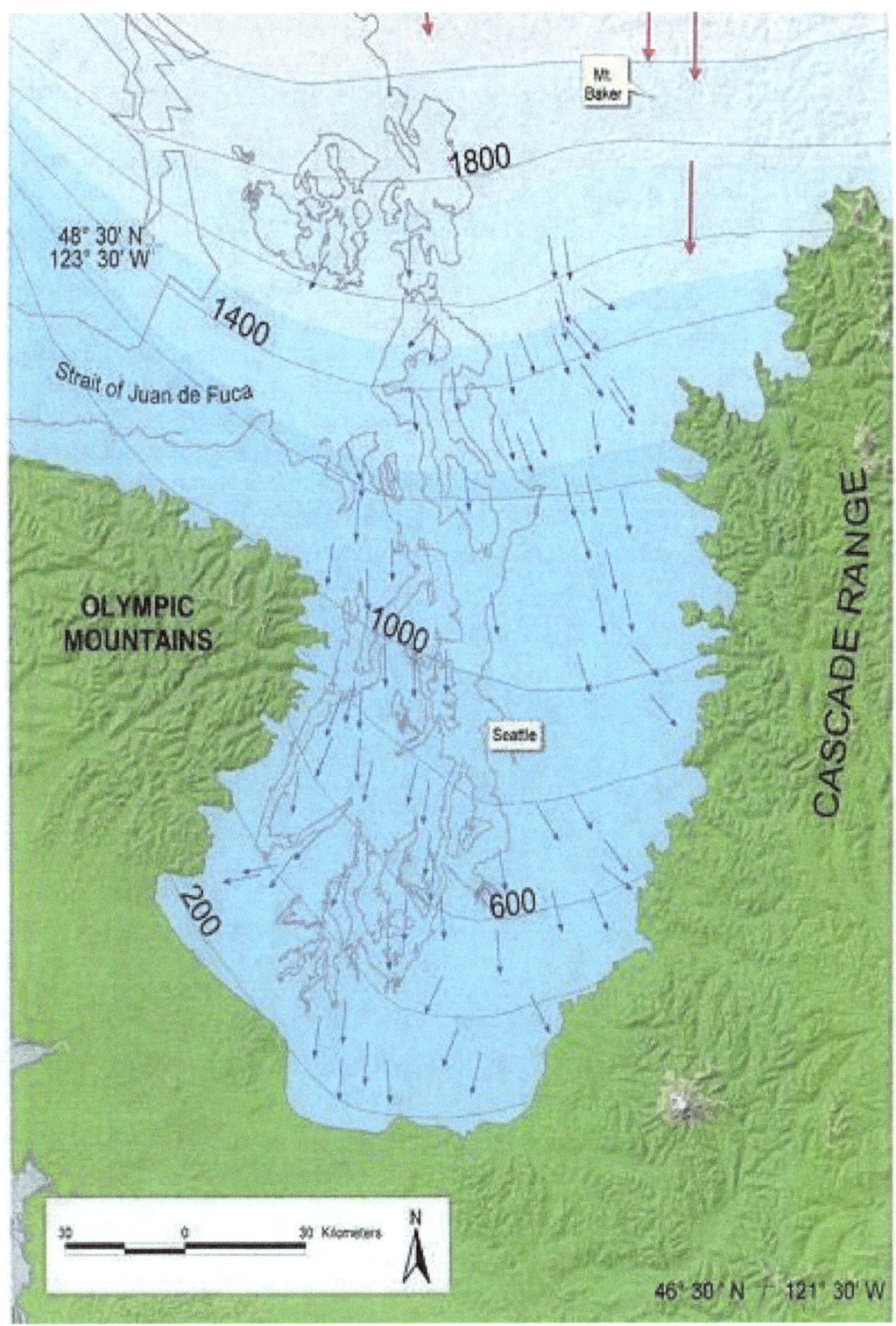

Figure 126B. The Cordilleran Ice Sheet (blue color) was about 6,000 feet thick near Bellingham about 15-20,000 years ago.

MT. BAKER DURING THE LAST ICE AGE

The Ice Age began about 2.5 million years ago when global temperatures cooled and immense continental ice sheets more than 10,000 feet thick covered vast areas of northern North America (Fig. 126), Europe, and Eurasia. The Cordilleran Ice Sheet advanced into Washington from British Columbia at least six times (and perhaps more), leaving a footprint of scoured land surface and glacial debris deposited by the ice. The enormous size of the ice sheet, which filled the Puget Lowland from the Olympic Mts. to the Cascades and from the Canadian boundary to Olympia, dwarfed the small alpine glaciers in the Cascade Range today.

During the Ice Age, the Cordilleran Ice Sheet overran the North Cascades around Mt. Baker, covering all of the peaks in the North Cascades below about 6,000 feet (Fig. 127). Glacial grooving and polishing of bedrock across high drainage divides (Fig. 128) show that the Cordilleran Ice Sheet flowed essentially north–south regardless of the underlying topography. Glacial erratics found at elevations of 5,700 feet (Fig. 129) indicate that the ice sheet must have been at least that high.

Figure 127. Reconstruction of the Cordilleran Ice Sheet in the Mt. Baker area during the last Ice Age. Gray color is the ice sheet.

Figure 128. Glacial grooves and polished rocks made by the Cordilleran Ice Sheet at Artist Point.

Figure 129. Glacial erratic from Mt. Herman on the summit of Table Mt.

THE NOOKSACK ALPINE GLACIER SYSTEM (NAGS)

The surface of the Cordilleran Ice Sheet at its maximum extent about 17,000 years ago was at least 6,000 feet above sea level and only peaks in the North Cascades above that elevation stood above the surface of the ice. Rapid melting of the ice sheet between 17,000 and 14,000 ^{14}C years ago resulted in lowering of the ice surface below ridge crests in the Nooksack drainage of the North Cascades, and glacial activity thereafter became topographically controlled by ridges and valleys. The glaciers at that point were no longer connected to the Cordilleran Ice Sheet and valley glaciers in the upper Nooksack Valley were fed by glaciers on Mt. Baker, Mt. Shuksan, and the Twin Sisters Range (Figs. 130, 131). Remnants of the Cordilleran Ice Sheet persisted in the lowland to the west at that time but were separated from the Nooksack glaciers by several ridges higher than the surface of the ice sheet.

Figure 130A. The Nooksack Alpine Glacier System about 12,500 ^{14}C years ago.

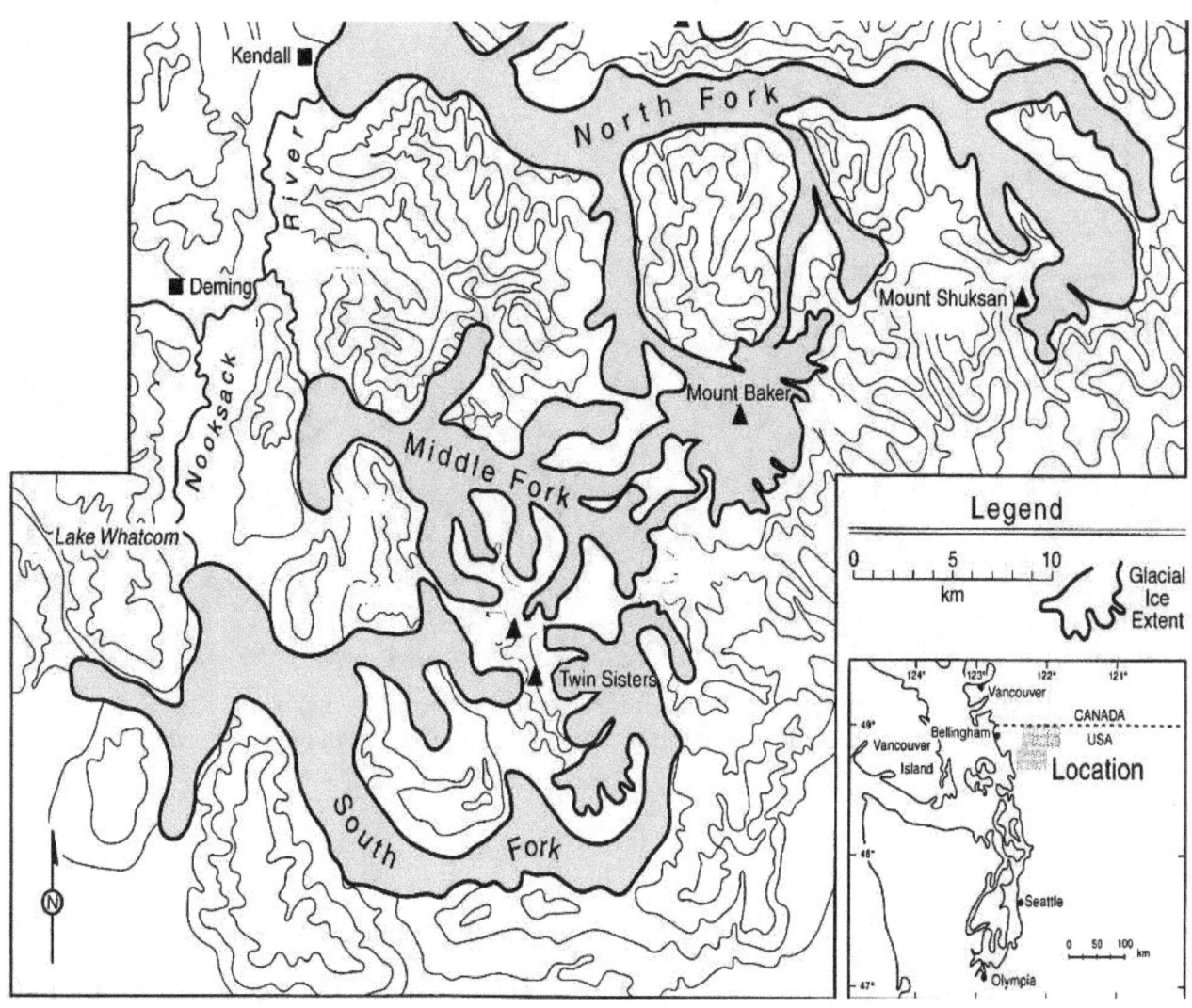

Figure 130B. The Nooksack Alpine Glacier System about 12,500 ^{14}C years ago. Gray areas are former glaciers.

Figure 131. Reconstruction of the Nooksack Alpine Glacier System (NAGS) about 12,500 ^{14}C years ago. Long valley glaciers extended from Mt. Baker and Mt. Shuksan down the Nooksack North Fork and Middle Forks.

These alpine glaciers made up the Nooksack Alpine Glacier System (NAGS), which deposited sediments in the Middle, North, and South forks of the Nooksack drainage 15–25 miles downvalley from their sources. The glacier split into three topographically controlled alpine glaciers (Fig. 130): (1) the North Fork glacier extended from Mt. Shuksan and Mt. Baker to Kendall; (2) the Middle Fork glacier reached the Nooksack South Fork and (3) the South Fork glacier extended to the southeast arm of Lake Whatcom and Cranberry Lake north of Sedro Woolley.

Nooksack Middle Fork Alpine Glaciers

The Mosquito Lake valley, a tributary of the Nooksack Middle Fork, has no stream in it and is filled with mounds and hollows composed of sand and gravel deposited from a stagnant glacier. Many of the depressions are up to 150 feet deep and contain lakes (Fig. 132).

An elongate ridge (lateral moraine) is draped across the irregular mounds and hollows near Mosquito Lake (Fig. 132). The moraine consists of 30 feet of poorly sorted glacial till containing numerous glacially faceted and striated boulders from Mt. Baker and the Twin Sisters Range (Fig. 132). The moraine is underlain by 60 feet of sand and gravel with many small faults and other collapse structures indicative of deposition against stagnant ice. Erratic boulders more than three feet in diameter on the moraine consist mostly of Mt. Baker lava (82.6%), Twin Sisters dunite (8.6%), Darrington phyllite, and Chuckanut sandstone, all of which make up the valley sides.

A 65–foot–deep peat bog north of Mosquito Lake occupies one of the deep depressions made by melting out of buried ice. Rootlets in gravel at the base of a core were dated at 12,356 ± 115 radiocarbon years, and wood on the gravel was dated at 12,165 ± 95 radiocarbon years. These dates indicate that the ice that once occupied the depression had melted away before 12,350 radiocarbon years ago (about 14,000 calendar years). The glacier later retreated eight miles up–valley from Mosquito Lake where it built a lateral moraine during the Younger Dryas cold period near the end of the last Ice Age that buried logs dated at 10,680 ± 70 and 10,500 ± 70 radiocarbon years ago.

Figure 132A. A long ridge (lateral moraine) draped over stagnant ice deposits and deep hollows filled with peat near Mosquito Lake. These deposits are slightly older than 12,350 radiocarbon years.

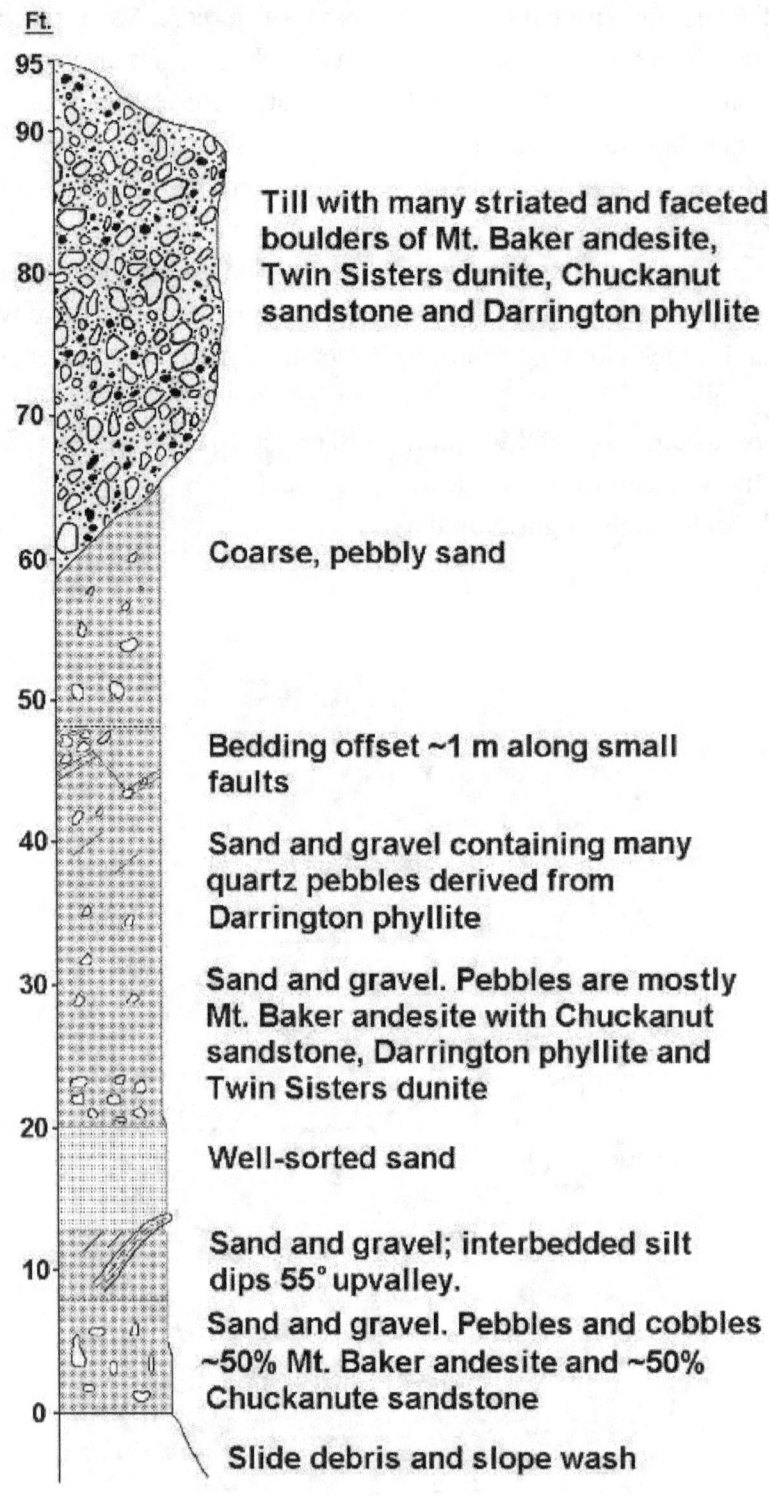

Figure 132B. Sediments composing the long lateral moraine near Mosquito Lake.

Figure 132C. Bouldery glacial till (top) making up the lateral moraine near Mosquito Lake.

North Fork Ice Age alpine glaciers

The Nooksack North Fork begins at a large cirque basin on the NW flank of Mt. Shuksan and trends westward in a long, broad, glacial trough (Fig. 131). Twenty five miles downvalley from the Mt. Shuksan cirque, remnants of two moraines are preserved along the north side of the valley (Fig. 133). The outermost of the two moraines, the Kendall moraine, makes a ridge several hundred yards long just upvalley from Kendall and the Maple Falls moraine lies a mile upvalley.

Pebbles and cobbles in the Kendall moraine consist of rock types from upvalley, mostly Chuckanut sandstone (54%) and Mt. Baker lava (12%). The composition of stones in the Kendall moraine shows a distinct, local, upvalley source. The Mt. Baker lava is significant because it must have come from Mt. Baker via Glacier Creek, a tributary of the upper Nooksack near the town of Glacier. The moraine also contains cobbles from a unique outcrop of Paleozoic volcanic rock found only in the upper North Fork drainage. In contrast, deposits of the Cordilleran Ice Sheet consists mostly of granite. An unusually large number of stones in the Kendall moraine are glacially faceted, striated, and polished, many more than at other localities in the Pacific Northwest. The reason for so many glacially abraded stones appears to be the long transport distance from the source of the ice 25 miles upvalley.

As shown by the composition of pebbles and cobbles in the glacial till, ice must have flowed northward from Mt. Baker into the North Fork valley in the direction opposite to the earlier flow of the Cordilleran Ice Sheet, confirming that the Kendall moraine was built by a local alpine glacier rather than the Cordilleran Ice Sheet.

Figure 133A. Moraines of the Nooksack Alpine Glacial System in the Nooksack North Fork near Kendall.

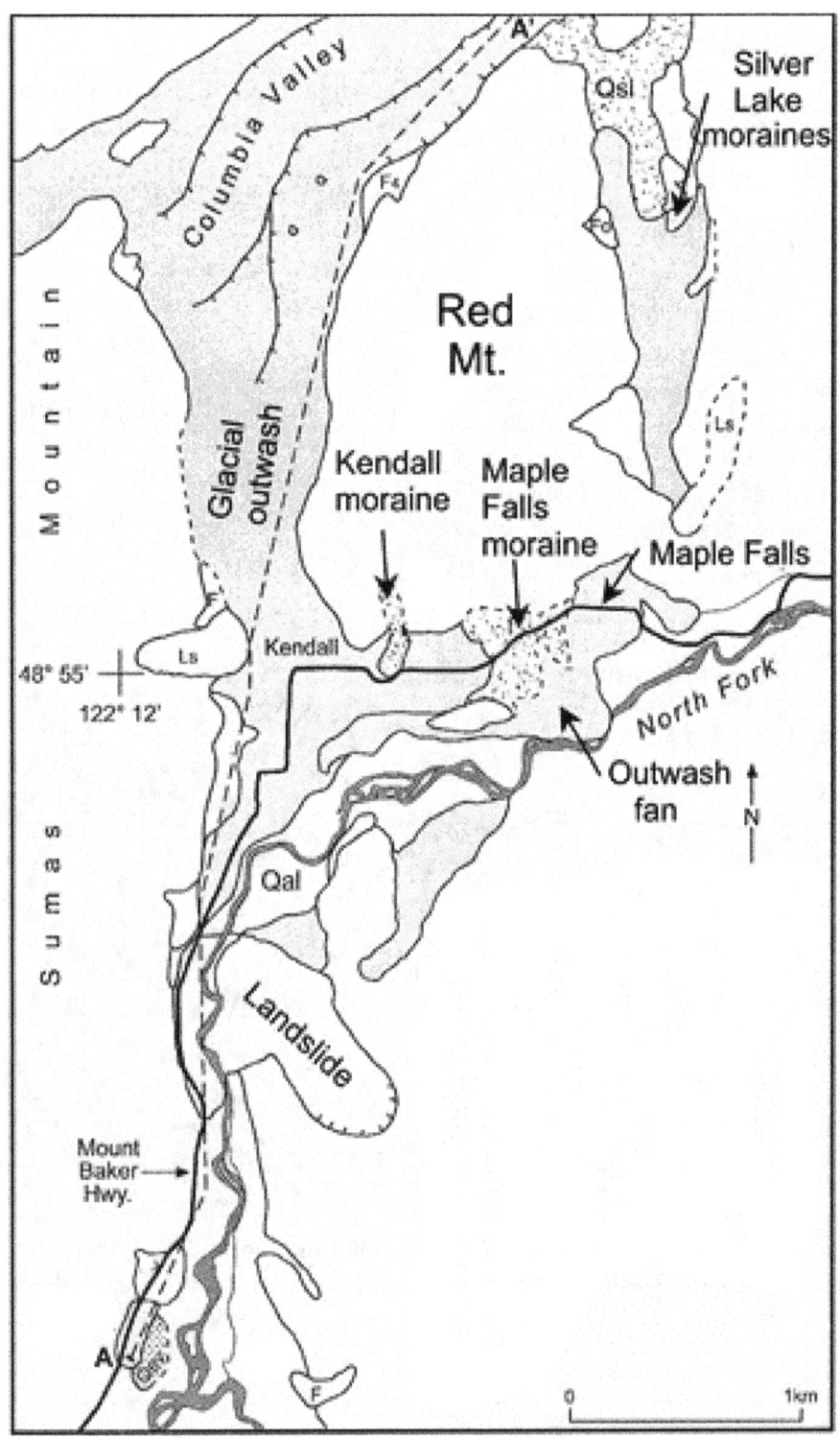

Figure 133B. Moraines of the Nooksack Alpine Glacial System in the Nooksack North Fork near Kendall.

South Fork alpine glaciers

A lateral moraine that makes a ridge near the east end of Lake Whatcom along the south shore of the lake was deposited by the South Fork alpine glacier, which extended 25 miles downvalley from the Twin Sisters Range (Fig. 134). The moraine is made up almost entirely of Darrington phyllite from the valley to the east and dunite from the Twin Sister Range. Many large, faceted, and striated dunite boulders derived from the Twin Sisters Range occur in till of the lateral moraine (Fig. 135). Of 109 boulders at least two feet in diameter exposed in a deep excavation one mile west of the east end of the lake, 91 were Darrington phyllite and 11 were Twin Sisters dunite, indicating that the glacier that deposited the moraine came from the Twin Sisters Range to the east, rather than from the Cordilleran Ice Sheet.

A peat bog about 35 feet deep at Cranberry Lake occurs at the terminus of the former South Fork glacier seven miles south of Wickersham. Basal bog dates (12,215 to 12,733 radiocarbon years), are generally equivalent to those at the deep bog in the Middle Fork valley near Mosquito Lake, suggesting that the glaciers in both valleys were the same age.

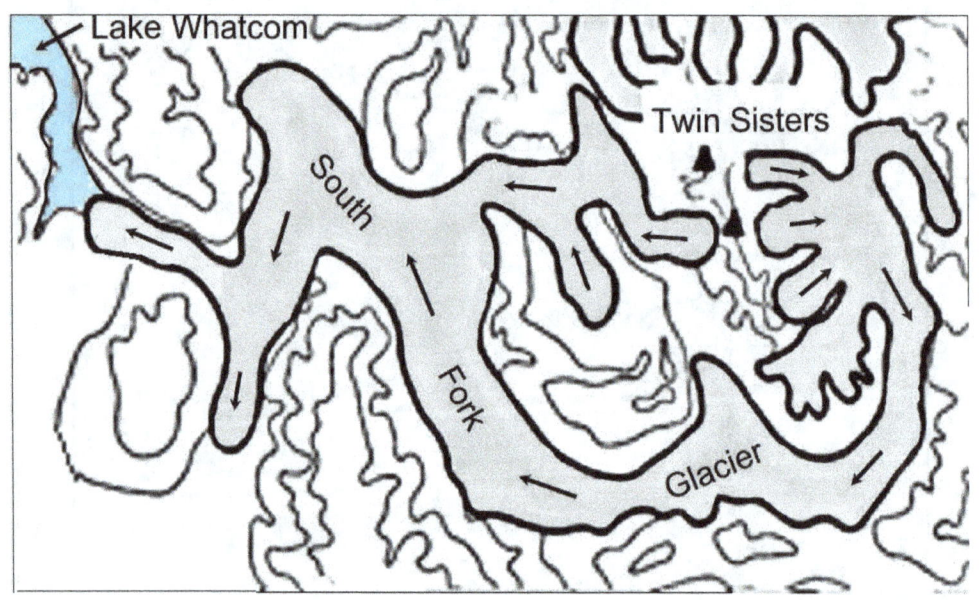

Figure 134. South Fork glacier of the Nooksack Alpine Glacial System. Ice flowed from the Twin Sisters Range into the eastern arm of Lake Whatcom.

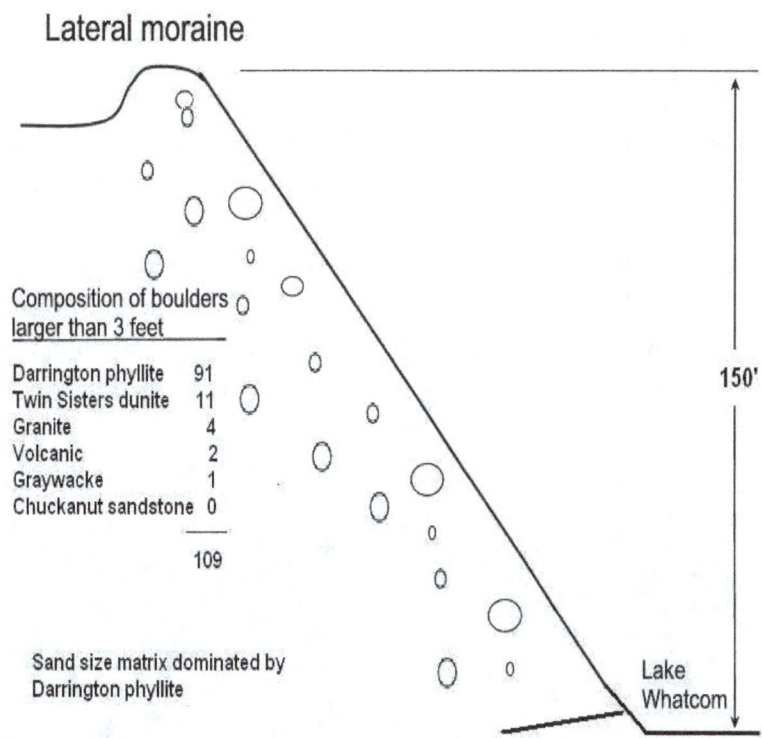

Figure 135. .[top] Cross section of the lateral moraine near the east end of Lake Whatcom. Boulders of dunite in the moraine proves that the glacier came from the Twin Sisters Range. [bottom] Glacial deposits of phyllite and dunite making up the lateral moraine.

GLACIATION OF HEATHER MEADOWS–ARTIST POINT

The scenery at Heather Meadows and Artist Point (Fig. 136) is a sweeping panorama of some of the most beautiful mountains in the world. To the east is Mt. Shuksan, (9,127') with hanging glaciers that break off in thundering avalanches all summer. To the southwest is the glacier clad, steaming Mt. Baker (10,785') volcano. To the west is Table Mt, a long ridge of lava that was once a valley, now indented by glacial cirques occupied by Bagley and Chain Lakes. Visible to the north are Tomyhoi Peak (7,451'), American Border Peak (8,068'), Red Mt. (also known as Mt. Larrabee) (7,868'), Winchester Mt. (6,521'), Goat Mt. (6,891'), and Mt, Sefrit (7,191'). To the south is Baker Lake with Glacier Peak on the far skyline.

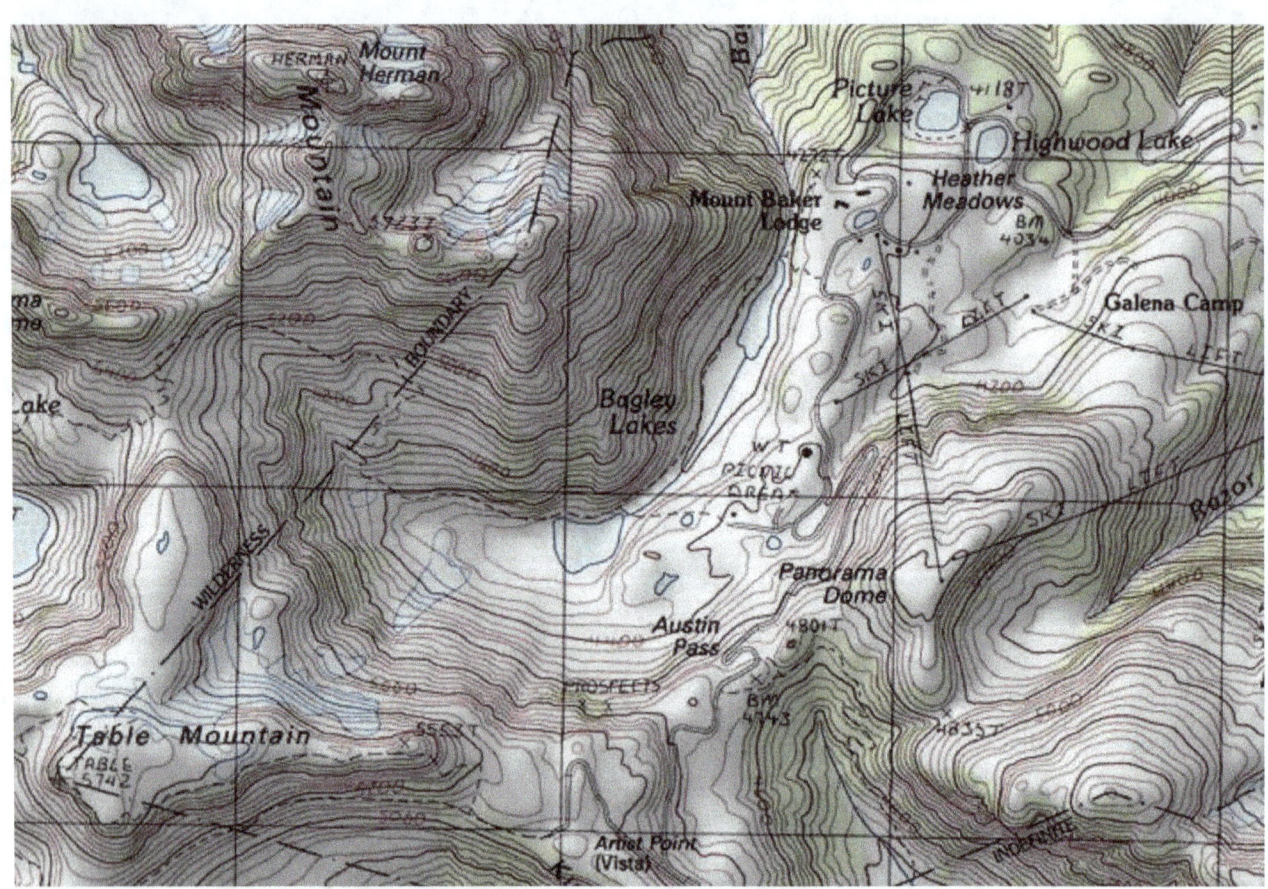

Figure 136. Map of Heather Meadows area. Bagley Lakes, Picture Lake, and Highwood Lake were all made by alpine glaciers.

Picture Lake–Highwood Lake Moraines

At its maximum, the Cordilleran Ice Sheet covered all of the peaks below 6,000 feet in the North Cascades near Mt. Baker. Global temperatures warmed abruptly about 15,000 years ago and ice sheets all over the world melted suddenly. The Cordilleran Ice Sheet shrank rapidly and disappeared from the North Cascades, allowing local alpine glaciers to occupy cirques and extend downvalley. Alpine ice filled the cirque basins now occupied by Bagley, Picture, Highwood, and Chain Lakes.

Figure 137. Moraine holding in Picture Lake, Heather Meadows.

The alpine glacier that occupied the cirque basin of Bagley–Picture Lakes (Fig. 139A) deposited moraines mantled with 6,850 year old Mazama ash, which can be seen in the roadcut across the road from the Picture Lake trail entrance (Fig. 138). The trail circles the lake and offers spectacular views of Mt. Shuksan reflected in the lake (Fig. 137).

Figure 138. Mazama ash lying on the glacial moraine holding in Picture Lake.

Heather Meadows

Figure 139A shows a reconstruction of the former glacier in the Bagley Lakes–Picture Lake basin. The glacier flowed from the cirque at Table Mt. to Picture and Highwood Lakes (Figs. 136, 140-142) where it built moraines that hold in the lakes. A radiocarbon age of 9,410 ± 60 radiocarbon years (about 11,000 calendar years) before present was obtained from basal peat in Highwood Lake, indicating that ice had retreated from the moraine by then and that the moraine must be older than 9,400 radiocarbon years.

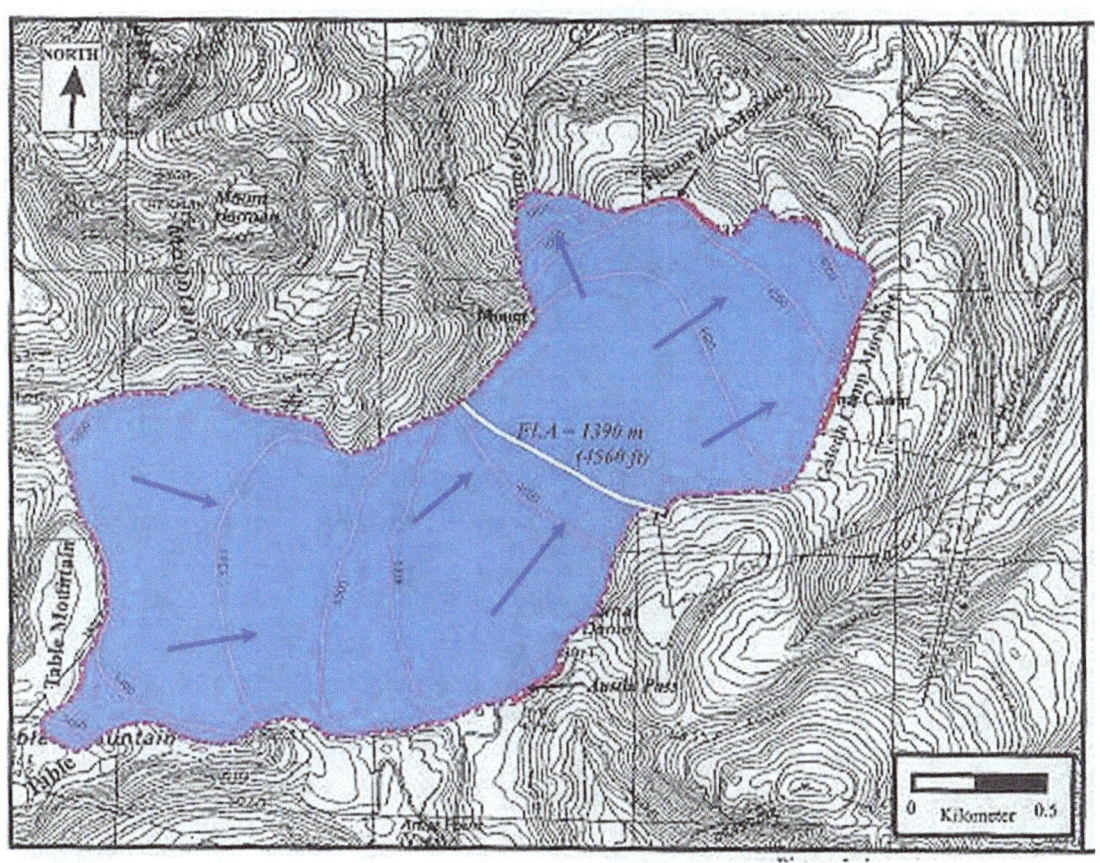

Figure 139A. Reconstruction of glacier occupying the Bagley Lake–Picture Lake area at Heather Meadows. (Burrows, 2001)

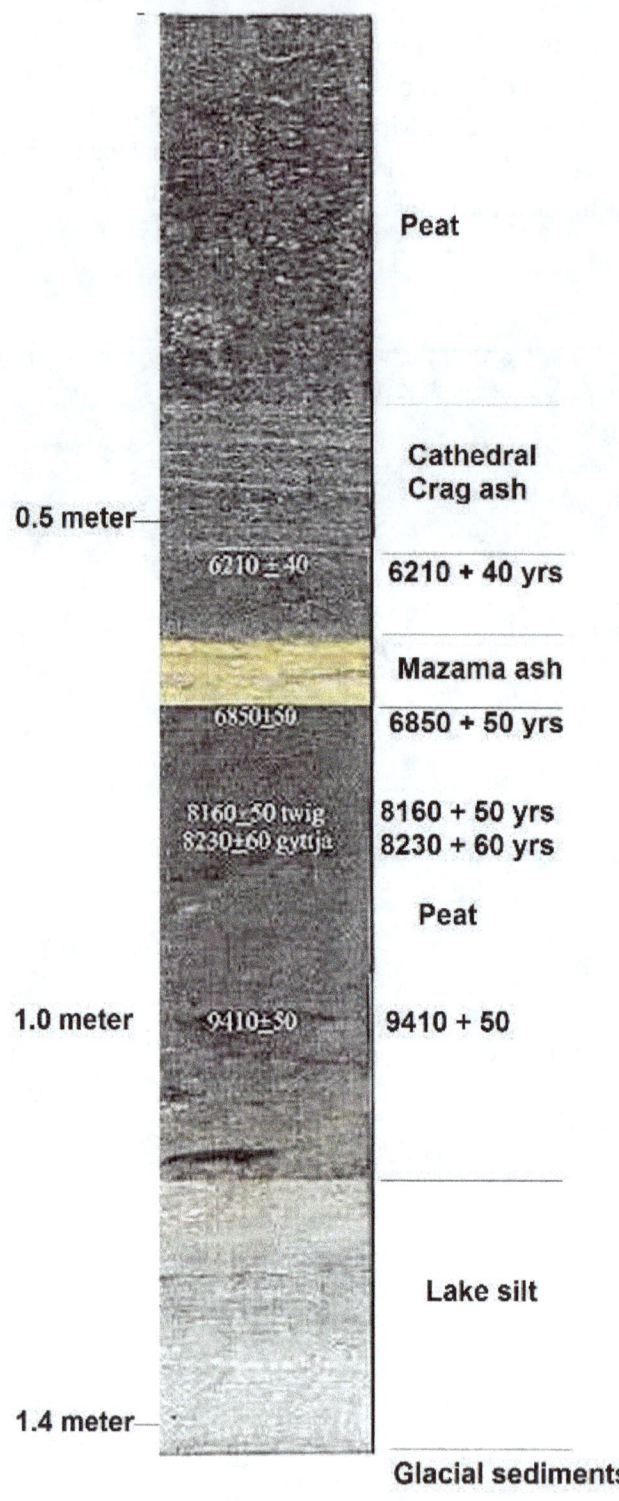

Figure 139B. Sediment core of Highwood Lake. (Modified from Burrows, 2001)

Bagley and Chain Lakes

Bagley and Chain Lakes lie on the floors of glacial cirque (Figs. 136, 140, 141) that were eroded by glaciers (Fig. 139). Iceberg Lake (Fig. 142) on the opposite side of Table Mt. also lies on a cirque floor made by an alpine glacier near the end of the last Ice Age about 11,000 years ago.

Figure 140A. Bagley Lakes from Austin Pass. The alpine glacier that made the lakes extended to the far right end of the photo.

Figure 140B. Upper Bagley Lake.

Figure 141. Upper Bagley Lake in a glacially eroded basin (cirque) cut into Table Mt. The cliffs beyond the lake were formed at the head of the glacier that occupied the basin.

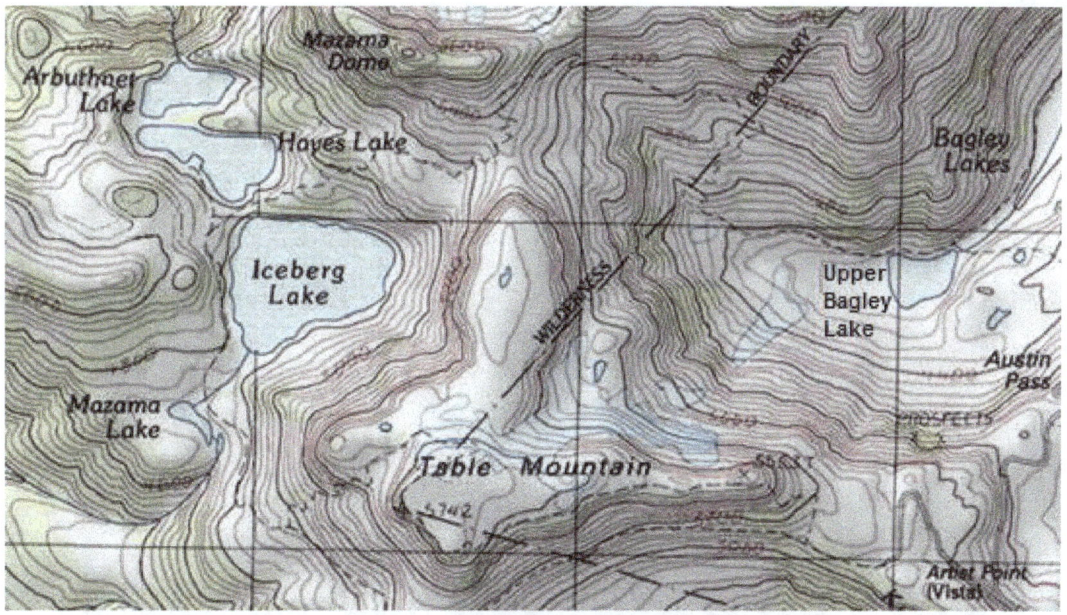

Figure 142. Bagley and Iceberg Lakes in cirques carved into Table Mt. during the last Ice Age.

Other examples of alpine cirques and moraines of last Ice Age (11,000 years ago) include Park Butte (Figs. 143, 144) and Pocket Lake (Figs. 143, 145, 146) above Schreibers Meadow on the SW flank of Mt. Baker. Many other glacial basins in the North Cascades contain similar moraines.

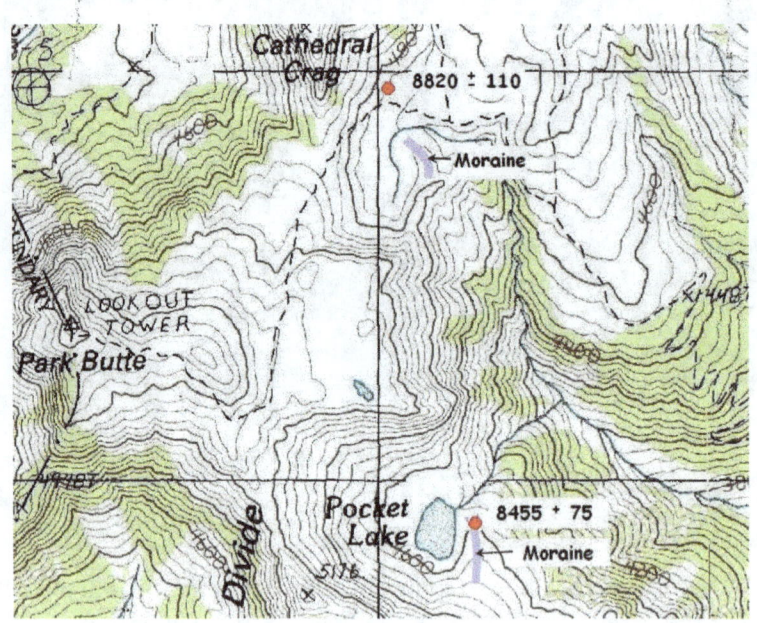

Figure 143. Park Butte and Pocket Lake cirque moraines near Park Butte.

Figure 144. Park Butte moraine and cirque, upper meadow above Schreibers Meadow.

Figure 145. Pocket Lake in a glacial basin (cirque) above Schreibers Meadow. The moraine holding in the lake contains organic material dated at 8,400 radiocarbon years.

Figure 146. Pocket Lake moraine containing wood dated at 8,400 radiocarbon years. The inner core of the moraine was probably built during the Younger Dryas cold period (about 11,000 years ago) and the outer part during the early Holocene (8,400 ^{14}C years ago).

Near the end of the last Ice age, (11,000 years ago), the Cordilleran Ice Sheet melted away, leaving alpine glaciers downvalley from Mt. Baker and Mt. Shuksan and in North Cascade cirques. These alpine glaciers melted away during the warm period that followed. Isotope analyses of Greenland and Antarctica show that from 10,000 to 1,500 years ago, global temperatures were warmer than present, but then turned cooler and the Earth entered the Little Ice Age. Alpine glaciers once again advanced but were not as extensive as those 11,000 years ago.

THE LITTLE ICE AGE (1300 A.D. TO THE 20TH CENTURY)

During the Medieval Warm Period (900 AD to 1300 AD), global temperatures were warmer than at present and human populations thrived. The Vikings colonized Greenland, wine was produced in latitudes north of present limits, food was relatively plentiful, and civilization flourished. At the end of Medieval Warm Period, about 1300 AD, temperatures suddenly dropped several degrees in about 20 years at high latitudes, and the Little Ice Age began. The cold climate was devastating to many regions, especially Europe and other areas at mid to high latitudes. The principal food supply during the Medieval Warm Period was cereal grains, and the colder, climate, early snows, shorter growing season, violent storms, and recurrent flooding swept Europe, causing massive crop failures. Grain crops that had been the main source of food for generations failed repeatedly. About one third of the population of Europe perished from famine and disease. Winters were bitterly cold in many parts of the world and glaciers advanced worldwide. Glacial advances in the European Alps in the mid–17th century encroached on farms and villages. Pack ice extended southward in the North Atlantic, surrounding Iceland and extending for miles in every direction, closing many harbors. The population of Iceland decreased by half. The Thames River in London froze over, and Viking colonies in Greenland died out because they could no longer grow enough food. Warm weather crops that had been grown for centuries in China were abandoned. Early settlers in North America, experienced exceptionally severe winters. All of the glaciers on Mt. Baker advanced and left a record in glacial deposits far downvalley.

The world has been slowly thawing out since the Little Ice Age but warming has not been continuous. Numerous 25-35 year warm/cool periods are apparent in glacial fluctuations and in isotope records in Greenland ice cores. These climate changes affected all of the glaciers on Mt. Baker.

MODERN GLACIERS

Coleman and Roosevelt glaciers

The Coleman glacier is named after Edmund Coleman, an Englishman from Victoria, B.C. He and Edward Eldridge, John Tennant, David Ogilvy, and Thomas Stratton were the first to climb Mt. Baker, August 17, 1868. The Coleman glacier originates at the saddle between the main summit cone of Mt. Baker and the Black Buttes and flows down the Glacier Creek valley (Figs. 147-150). The glacier terminus fluctuates back and forth as the climate changes, each time leaving a ridge of debris at the ice margin (moraine) or leaving a trim line in the adjacent forest.

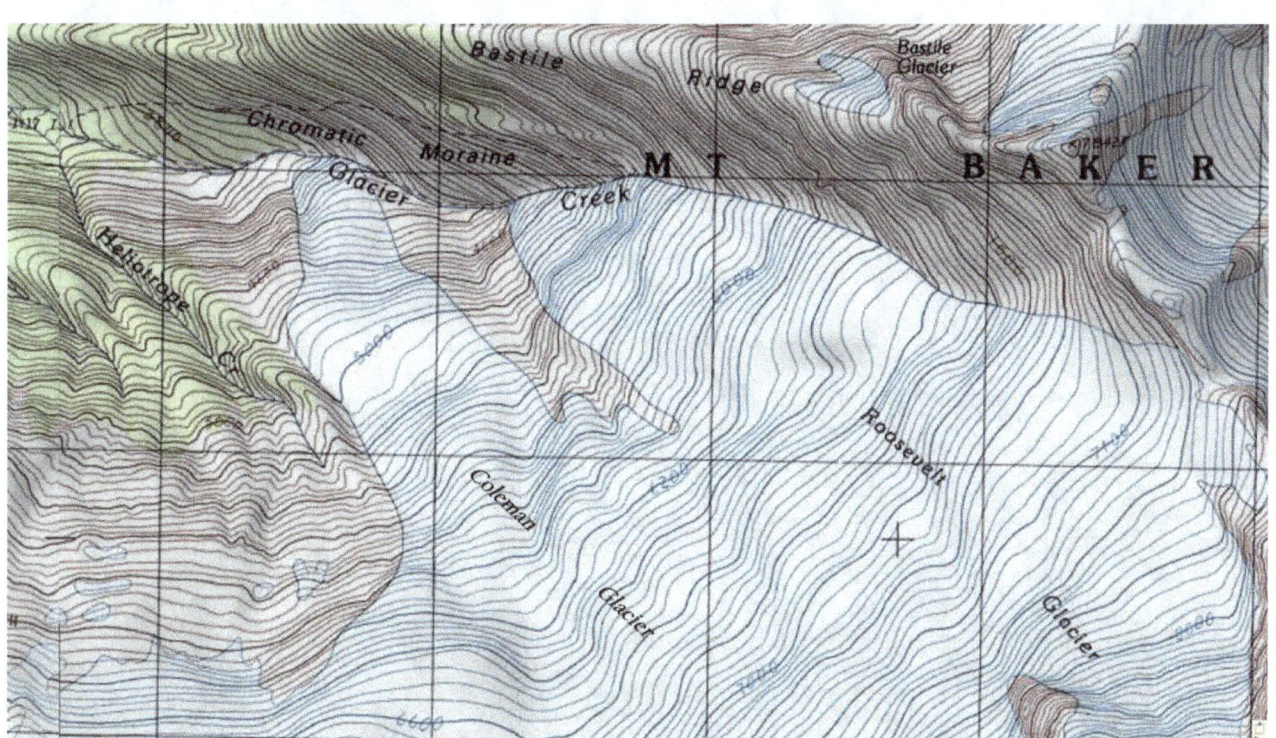

Figure 147. Coleman and Roosevelt glaciers.

Figure 148. Coleman and Roosevelt glaciers.

Figure 149. Terminus of the Roosevelt Glacier.

Figure 150. Terminus of the Coleman glacier.

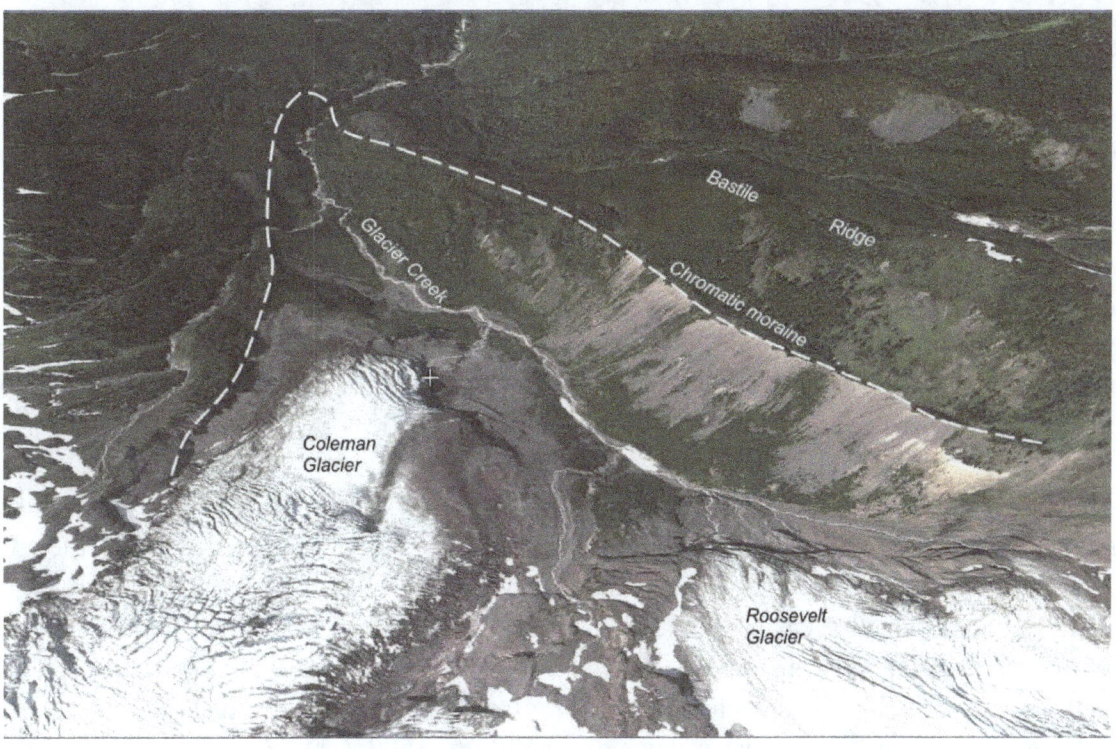

Figure 151. Former extent of the Coleman and Roosevelt glaciers during the Little Ice Age (about 1500 AD).

Little Ice Age (1300 AD to about 1915 AD)

During the Little Ice Age, the Coleman and Roosevelt glaciers coalesced and extended about 1¾ miles downvalley from the present terminus, (Fig. 151). The margins of this former glacier are distinctly evident in lateral moraines along the valley sides (Figs. 151, 152).

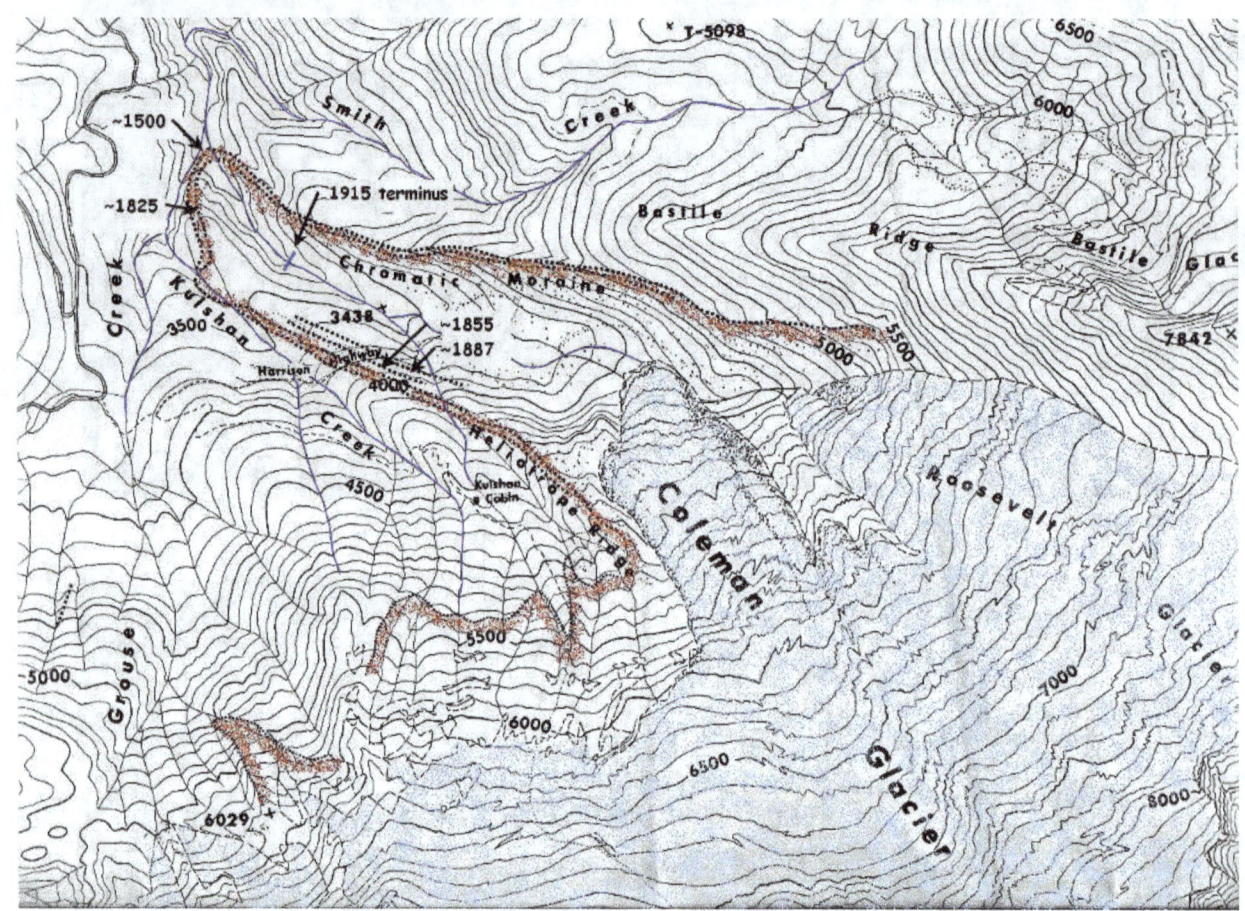

Figure 152. Former ice extent of the Coleman–Roosevelt glacier during the Little Ice Age about 1500 AD (brown pattern). (Modified from USGS topographic map)

A buried forest is exposed in the lateral moraine along the southwest side of Glacier Creek valley (Figs. 153-154). Large logs that protrude from the moraine are easily visible from the top of the moraine and may be seen by walking from the old Kulshan cabin site down the crest of Heliotrope Ridge. Radiocarbon dates of 680 ± 80 and 740 ± 80 radiocarbon years were obtained from the logs. That means the lower part of the lateral moraine was deposited during the cold period that just preceded the Medieval Warm Period. The trees grew on the lateral moraines during the Medieval Warm Period ending about 1300 A.D. but were later buried by a stronger glacial advance during the Little Ice Age about 1300 to 1500 A.D. A moraine about 1 ¾ miles downvalley from the present terminus has trees growing on it dating back to the 1500's.

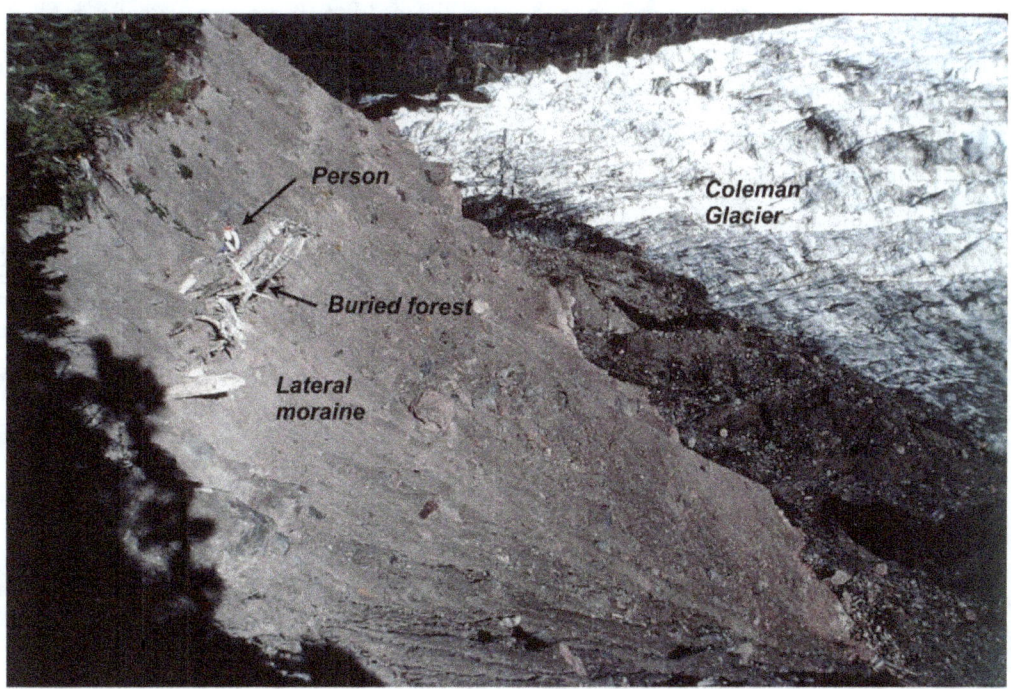

Figure 153. Medieval Warm Period forest buried by a lateral moraine of the Coleman glacier during the Little Ice Age.

Figure 154. Logs in a buried forest in the Coleman glacier lateral moraine, ^{14}C dated at 680 ± 80 and 740 ± 80 years

Successively younger moraines occur upvalley from the outermost moraine. Annual rings from trees growing on moraines show moraine–building episodes in the 1600s, ~1750, ~1790, ~1850, and ~1890. Each moraine represents a cold climate episode. The Little Ice Age glacial advance of the 1600s occurred in a cold period that took place during the Maunder Solar Minimum and the 1790 glacial advance occurred during the Dalton Solar Minimum 1790 to 1820. The 1890–1910 cool period, when many of the cold temperature records in North America were set, also occurred during a time of few sun spots.

During the global cold period from 1890 to about 1915, the Coleman glacier extended downvalley from its present terminus, almost all the way down to the 1500 AD Little Ice Age ice margin. This is documented by USGS topographic maps of 1909 and 1915 and a photograph taken at about the same time (Figs. 155-157).

Figure 155. 1909 map of the Coleman and Roosevelt glaciers which merged into a single glacier that extended far down valley from the present terminus.

Figure 156. 1915 topographic map of the Coleman- Roosevelt glacier system. The terminus of the glacier extended downvalley to an elevation of about 3,200 feet, almost to the position of the Little Ice Age terminus about 1500 AD.

Figure 157. The Coleman glacier early in the 20th century.

Fluctuations of the Coleman and Roosevelt glaciers from 1915 to 2015

Following the 1880 to 1915 cool period advance, the Coleman and Roosevelt glaciers retreated and advanced several times during the rest of the century. Water temperatures of the Pacific Ocean off the Washington coast fluctuate between warm and cool in a very regular pattern and Mt. Baker glaciers followed (Fig.158). When the ocean water was cool, Mt. Baker glaciers advanced and when the ocean was warm, the glaciers receded (Fig. 158).

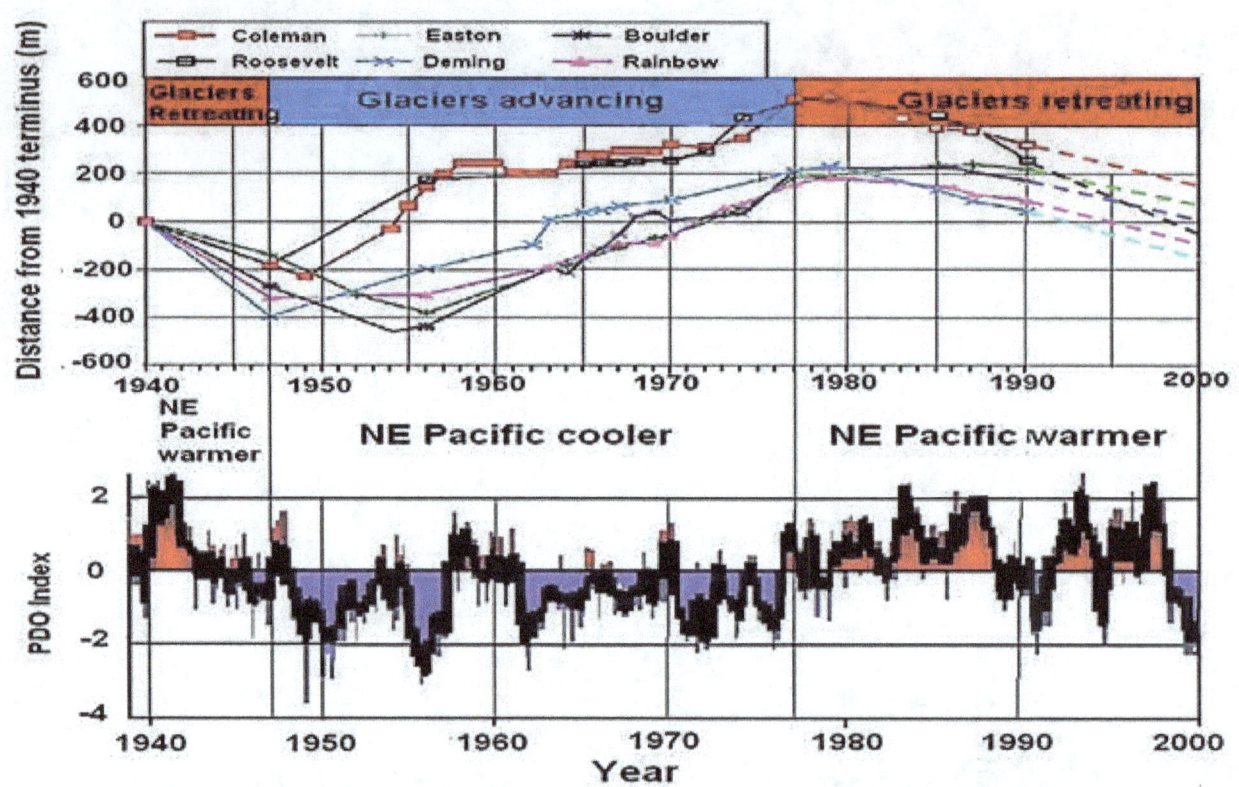

Figure 158. Fluctuations of Mt. Baker glaciers as Pacific Ocean water warmed and cooled. The same pattern occurs to the beginning of the century and for 500 years before that.

During the warm period from about 1915 to about 1945, termini of the Coleman and adjacent Roosevelt glaciers retreated rapidly upvalley. The Coleman glacier retreated about 1¾ miles from its 1915 margin to its 1940 position and retreated another 600 feet from 1940 to 1947.

The Roosevelt glacier retreated nearly two miles to its 1940 position and then retreated another 600 feet from 1940 to 1947. The average rate of recession of the termini of these glaciers was approximately 300 feet per year from 1915 to 1945.

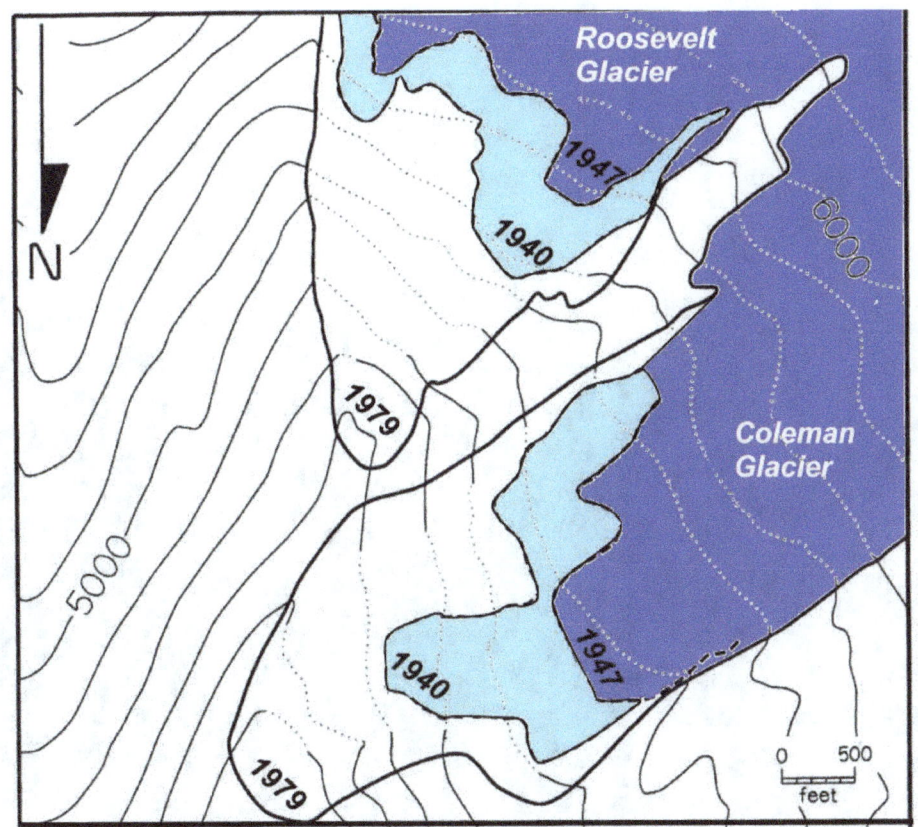

Figure 159. Historic fluctuations in the Coleman and Roosevelt glaciers. In 1915, the glacier terminus was more than a mile downvalley and retreated to the 1947 position during the ~1915 to ~1945 warm period then advanced well below the 1940 position by 1979. (Modified from Harper, 1992)

About 1945, the NE Pacific Ocean flipped from its warm phase into its cool phase, bringing to a close the 30–year glacial retreat and initiating a 30–year readvance of glaciers. Photographs of the period between 1915 and 1940 are rare, but beginning in 1940, many air photos are available to document the retreat. From 1947 to 1956, the Coleman glacier advanced 1073 feet (Figs. 160-162) and another 178 feet from 1956 to 1963. The Roosevelt glacier advanced 1175 feet from 1947 to 1956 and another 700 feet from 1956 to 1963 (Fig. 163). The average rate of advance of the glaciers from 1947 to 1956 was about 120–130 feet per year.

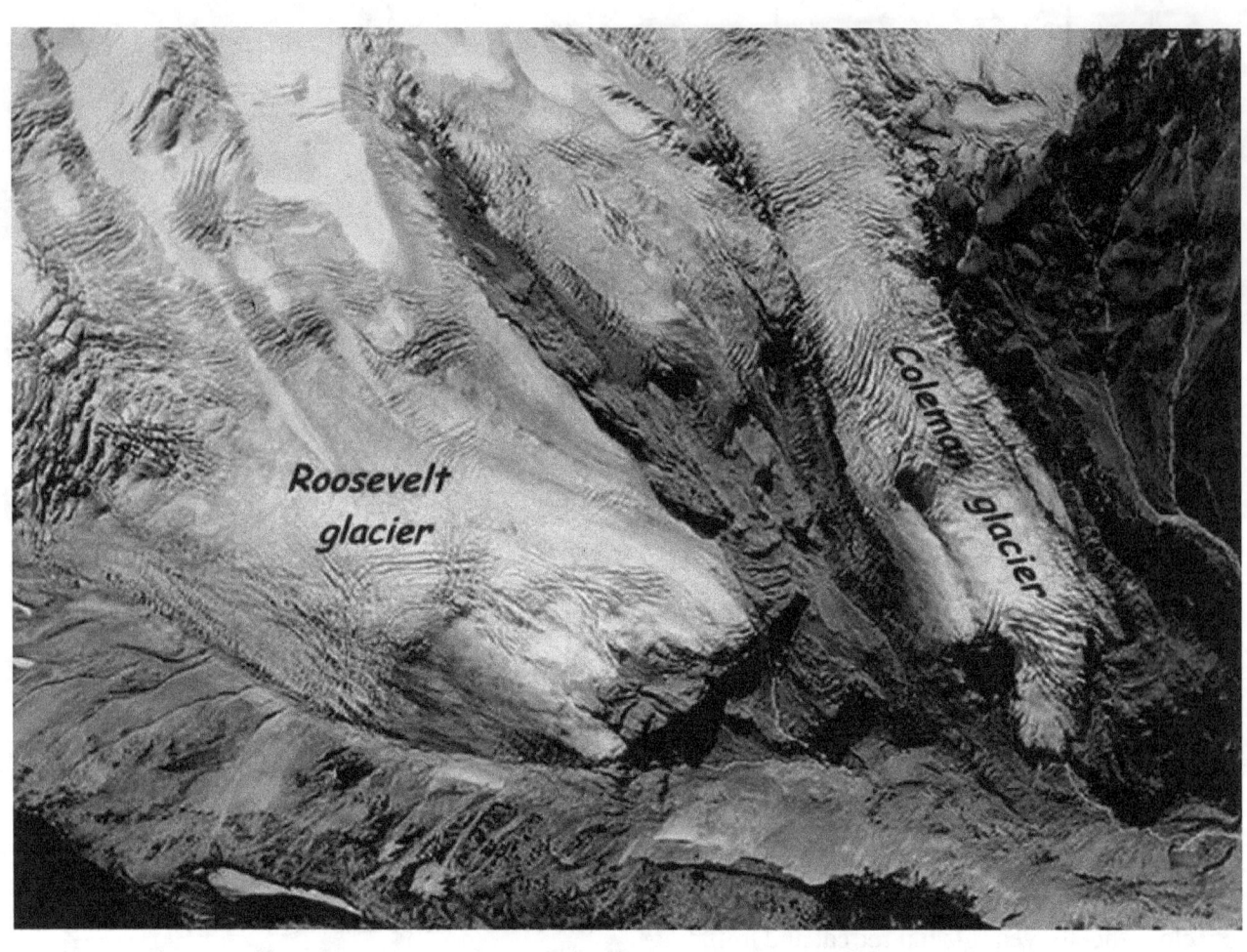

Figure 160. Roosevelt and Coleman glaciers in 1940. Note that the terminus of the Roosevelt glacier is just at the top of a dark cliff.

Figure 161. Roosevelt and Coleman glaciers in 1947. Note that the terminus of the Roosevelt glacier has receded back from the edge of the dark cliff where it was in 1940.

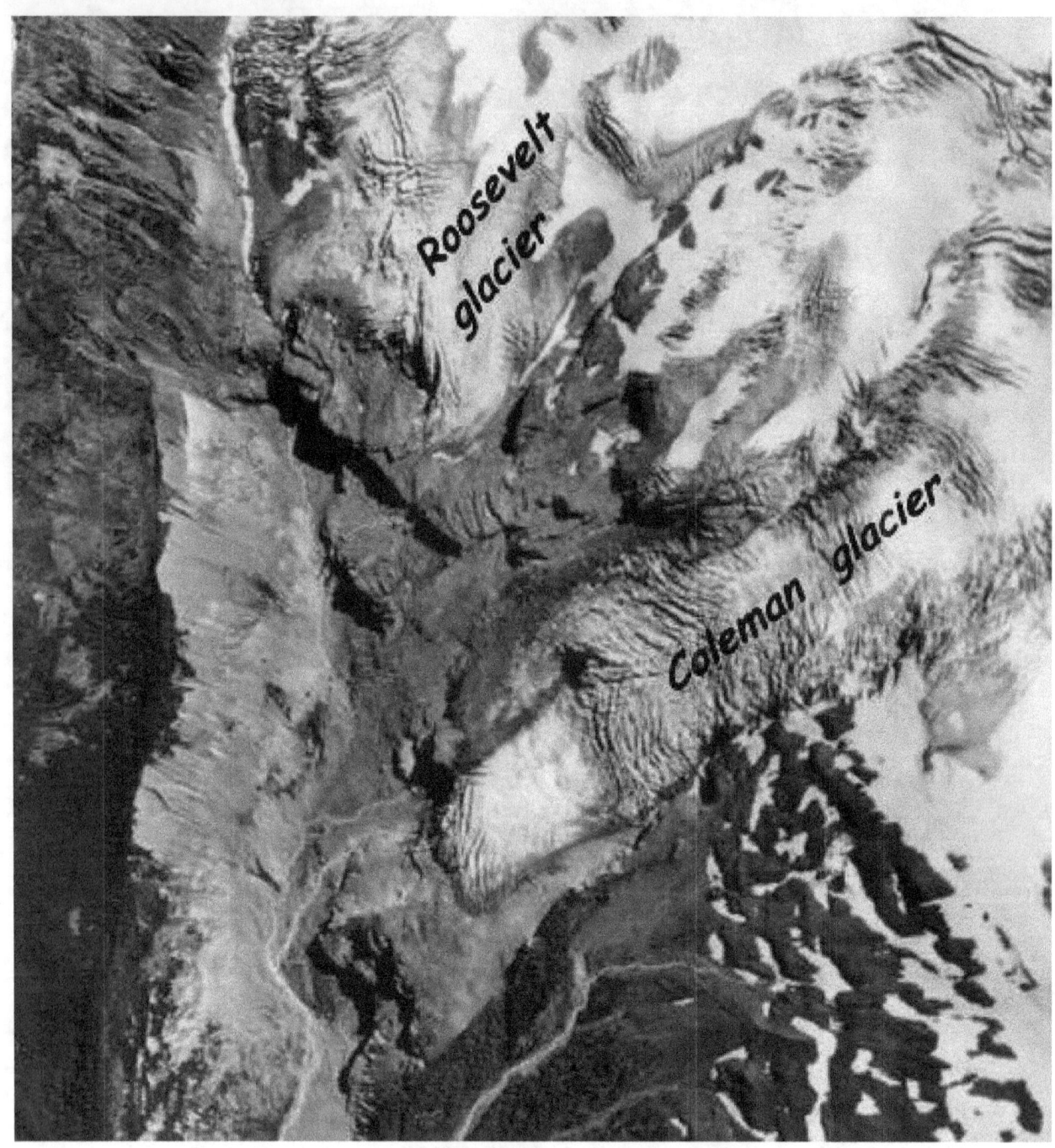

Figure 162. Roosevelt and Coleman glaciers in 1950

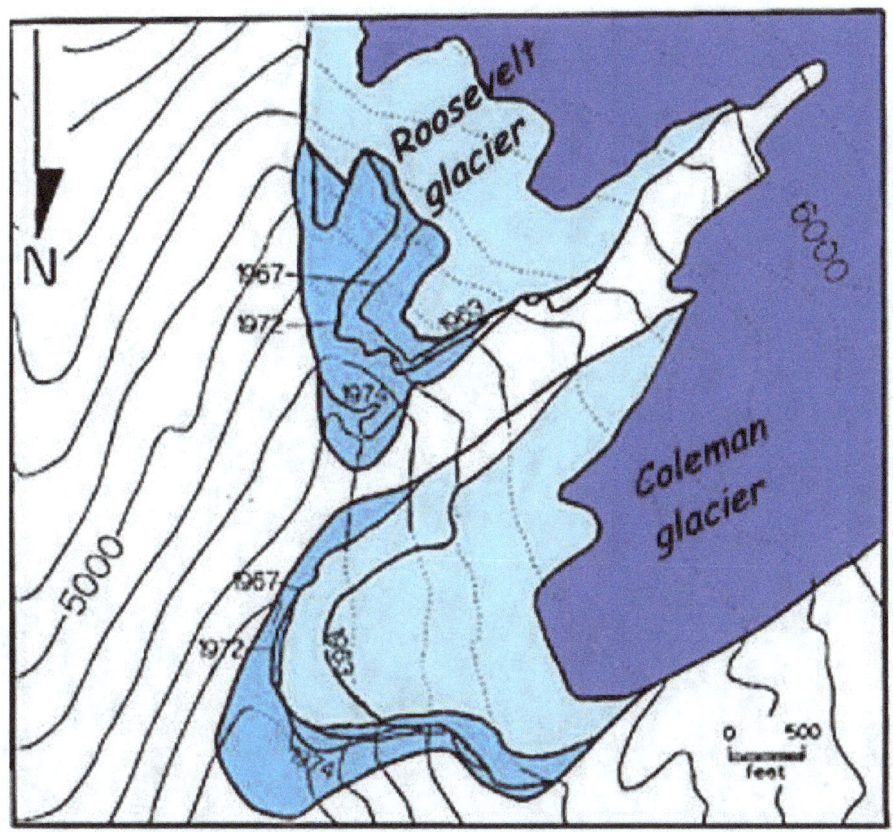

Figure 163. Terminal positions of the advancing Coleman and Roosevelt glaciers. (Modified from Harper 1992)

Figure 164. Ice-cored moraine at the Coleman glacier terminus.

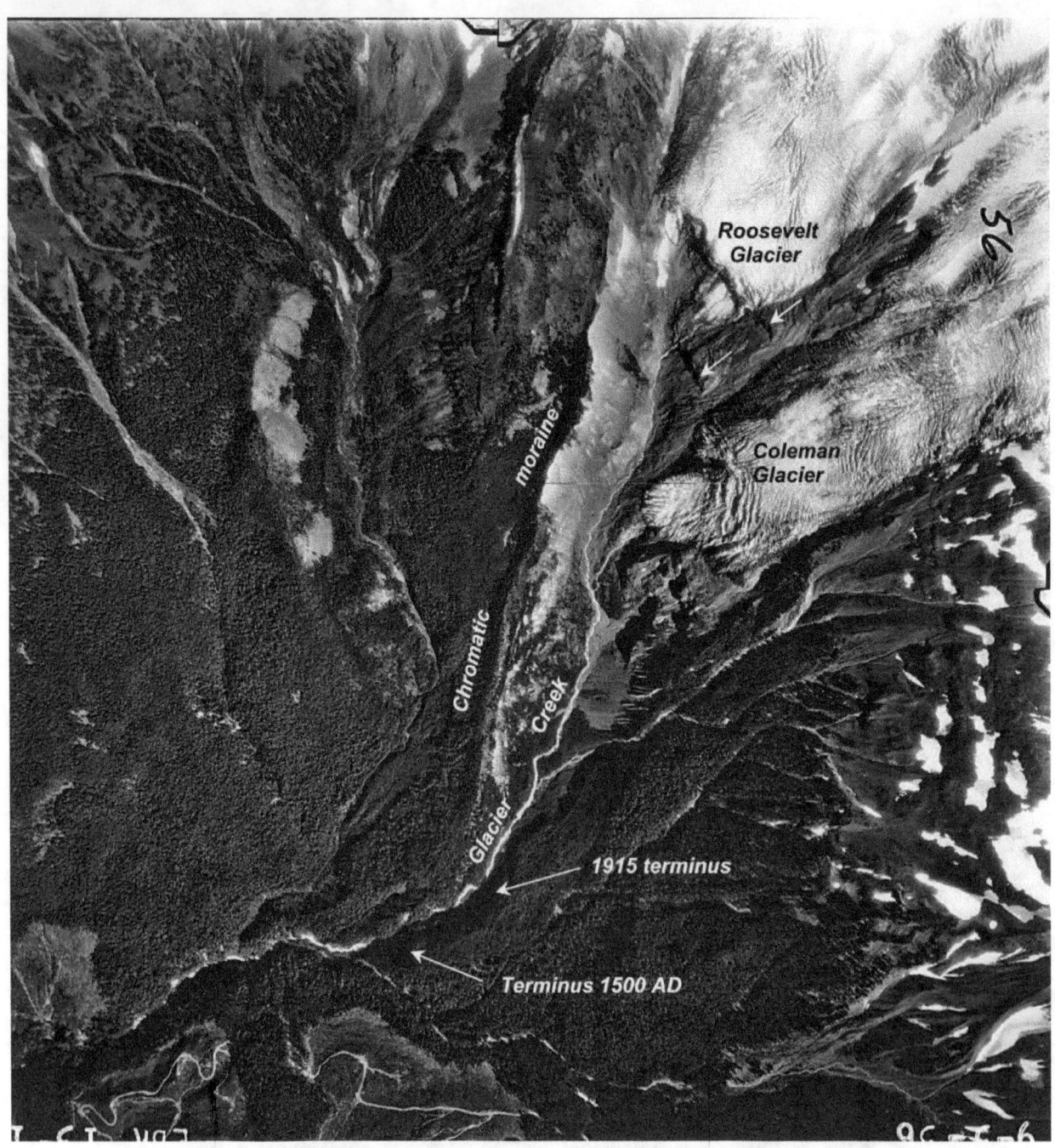

Figure 165. Coleman and Roosevelt glaciers, 1956. Note the two ledges just below the terminus of the Roosevelt glacier (arrows), which can be used as reference points for advance or retreat of the glacier. (Photo by Austin Post)

Figure 166. Coleman and Roosevelt glaciers, 1962. The Roosevelt glacier had spilled over the upper ledge (red arrow) but hadn't yet reached the lower ledge. (Photo by Austin Post)

From 1963 to 1967, The Coleman glacier advanced 254 feet and the Roosevelt glacier advanced 208 feet. The Coleman glacier advanced 140 feet from 1967 to 1970, 58 feet from 1970 to 1972, 128 feet from 1972 to 1974, 440 feet from 1974 to 1977, and 27 feet from 1977 to 1979, a total of 793 feet from 1967 to 1979, an average of about 66 feet per year. The Roosevelt glacier advanced 115 feet from 1967 to 1972, 472 feet from 1972 to 1974, 200 feet from

1974 to 1977, and 46 feet from 1977 to 1979 (Figs. 166-171). The highest rate of advance for the Coleman glacier was about 220 feet per year from 1972 to 1974 and the highest rate for the Roosevelt glacier was about 235 feet per year from 1972 to 1974.

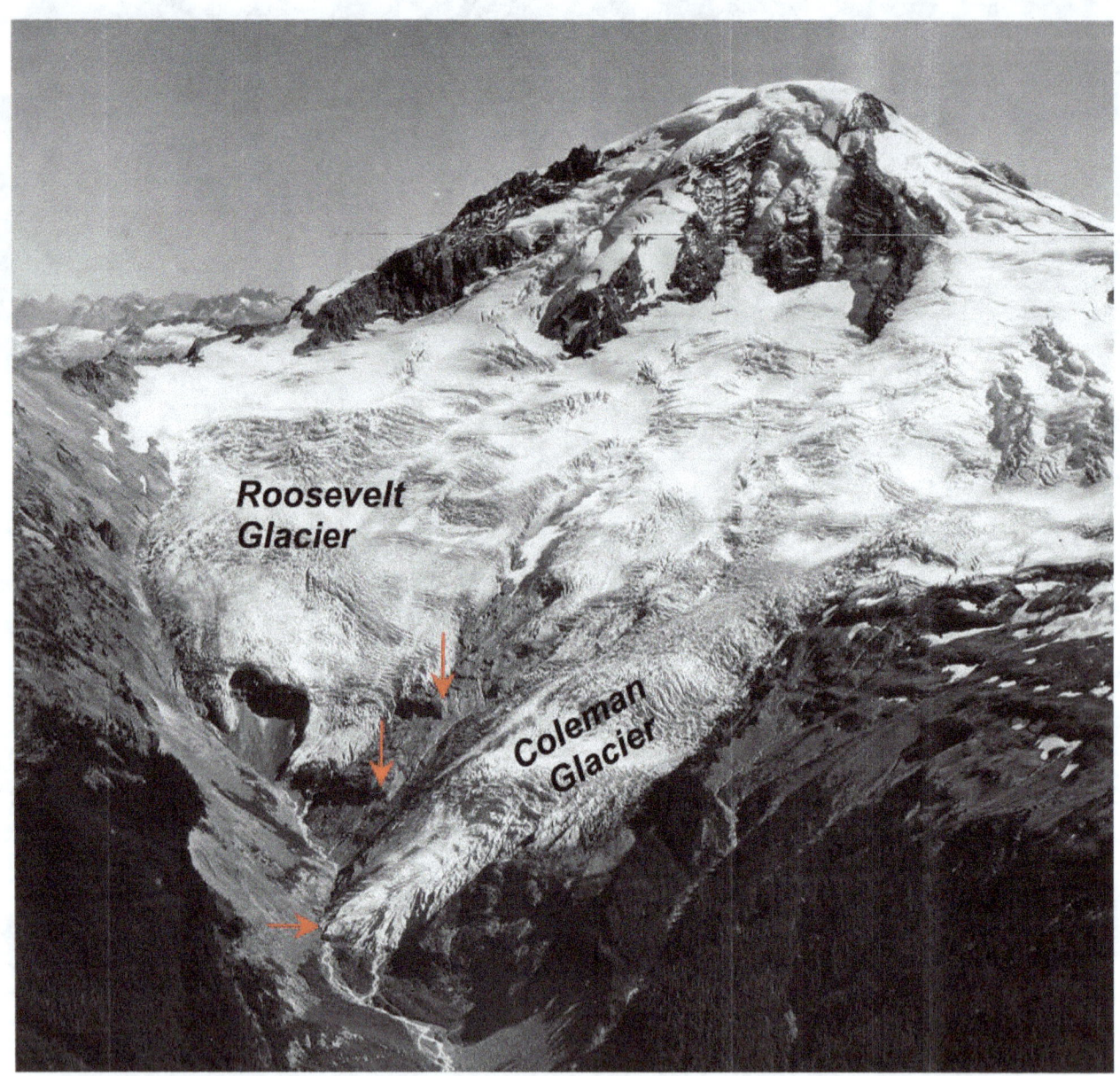

Figure 167. Coleman and Roosevelt glaciers, 1967. The Coleman reached across Glacier Creek (red arrow) and the Roosevelt extended to the lower of the two ledges below it (red arrows). (Photo by Austin Post)

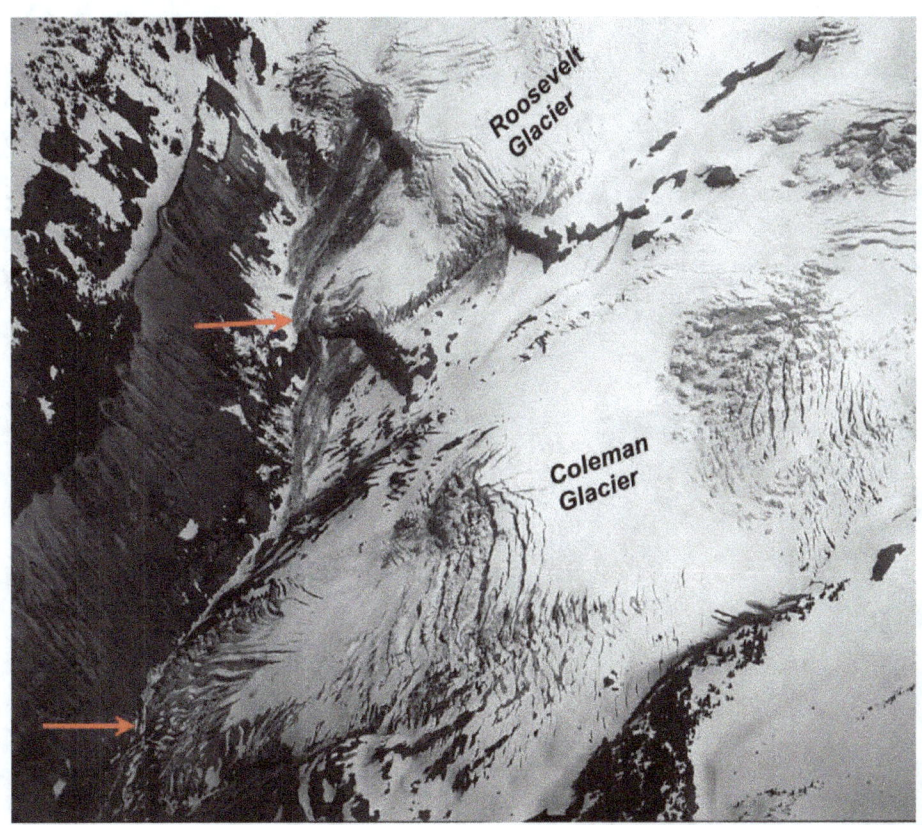

Figure 168. Coleman and Roosevelt glaciers, 1972. The termini of the Coleman and Roosevelt glaciers (red arrows) which have both crossed Glacier Creek. (Photo by Austin Post)

Figure 169. Roosevelt and Coleman glaciers nearly coalescing, 1975.

Figure 170. Coleman and Roosevelt glaciers, 1979. The Coleman Glacier extended across Glacier Creek and the Roosevelt glacier terminus nearly touched the Coleman glacier (red arrow). (Photo by Austin Post)

In 1977, the NE Pacific Ocean flipped abruptly back into its warm mode, the climate suddenly shifted back to warmer, and the glaciers once again began to retreat upvalley. From 1974 to 1977, the rate of advance of the Coleman glacier was 180 feet per year, but this dropped to 15 feet per year from 1977 to 1979. From 1979 to 1983, the Coleman glacier retreated 307 feet (77 feet per year). From 1979 to 1987, the Coleman glacier retreated 475 feet (60 feet per year).

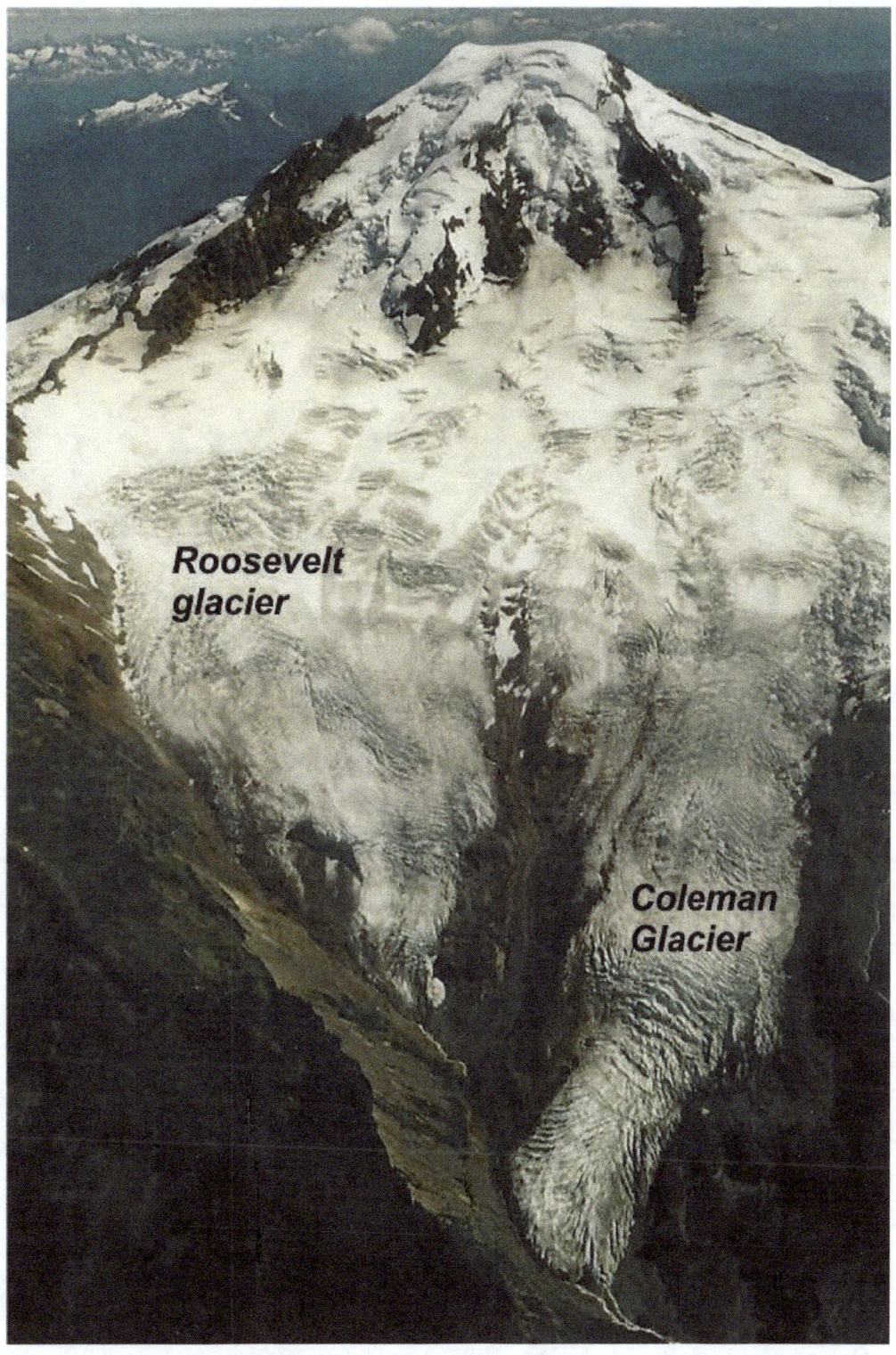

Figure 171. Coleman and Roosevelt glaciers, 1987. (Photo by Austin Post)

Figure 172. Coleman and Roosevelt glaciers, 1987. The Coleman glacier crosses Glacier Creek (red arrow) and the Roosevelt glacier has retreated upvalley (red arrow). (Photo by Austin Post)

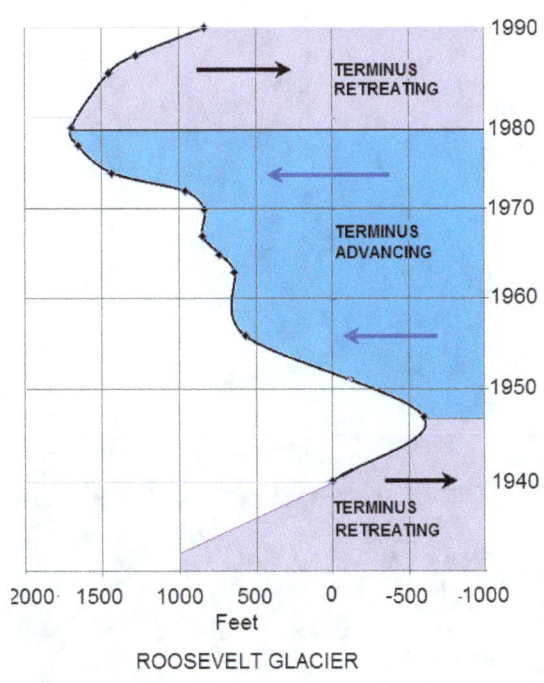
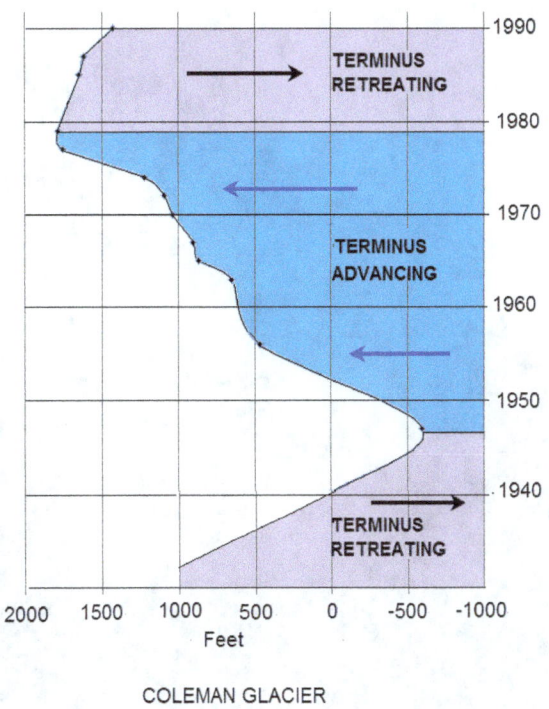

Figure 172. Graphs of the advance and retreat of the Roosevelt and Coleman glaciers from 1940 to 1990. Both glaciers retreated rapidly from 1915 to 1947, advanced strongly from 1947 to 1980, then retreated again from 1980 to 2015. The 2015 terminus hasn't yet retreated upvalley to its 1950 position.

Figure 173. Coleman and Roosevelt glaciers, 2002. The Coleman has retreated upslope from Glacier Creek and the Roosevelt glacier has retreated to the upper ledge. (Photo by Austin Post)

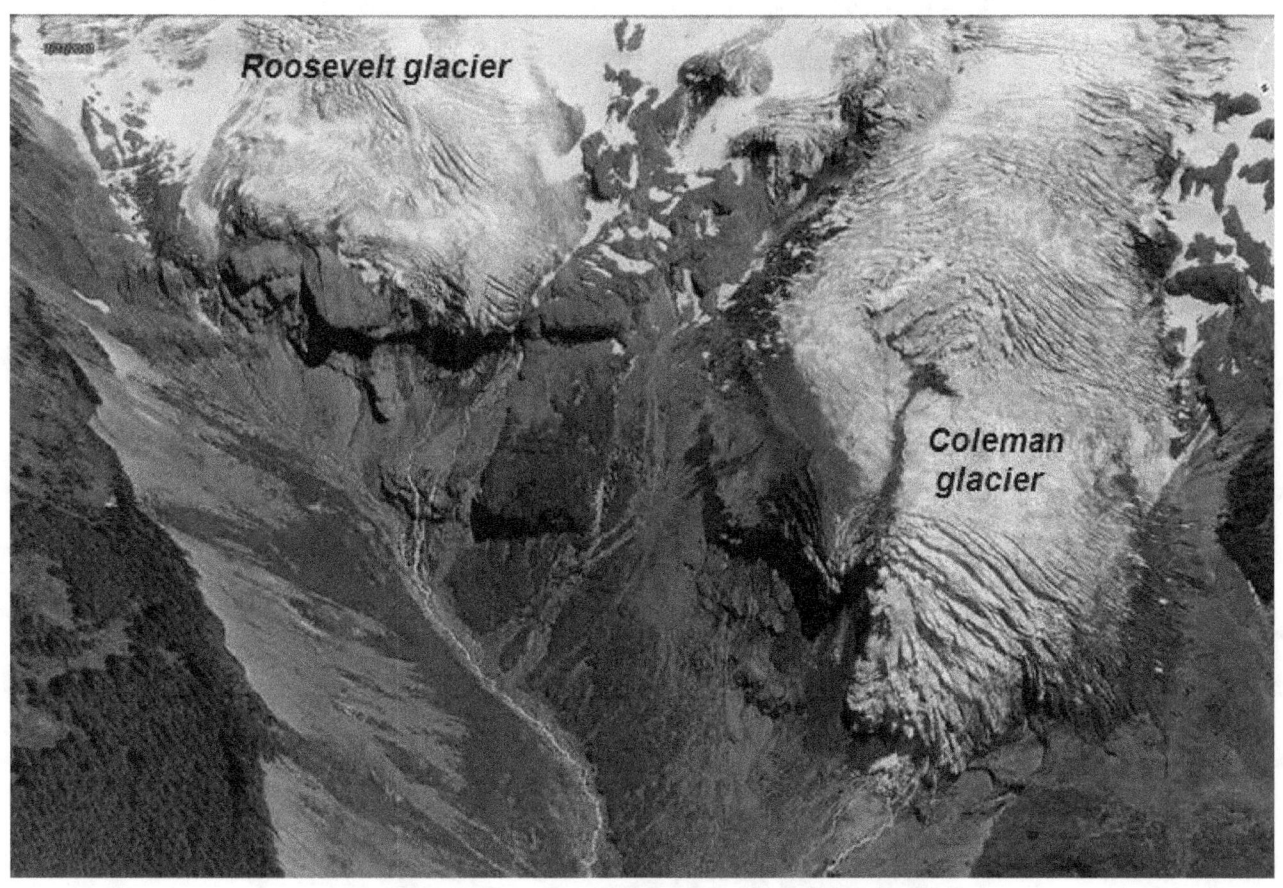

Figure 174. Coleman and Roosevelt glaciers, 2013.

Comparison of the position of the terminus of the Roosevelt glacier in 1947 and 2015 (Fig. 175) shows that both the Coleman and Roosevelt glaciers are more extensive now than they were in 1947. This is also shown on USGS topographic maps of 1952 and 2014 (Fig. 176).

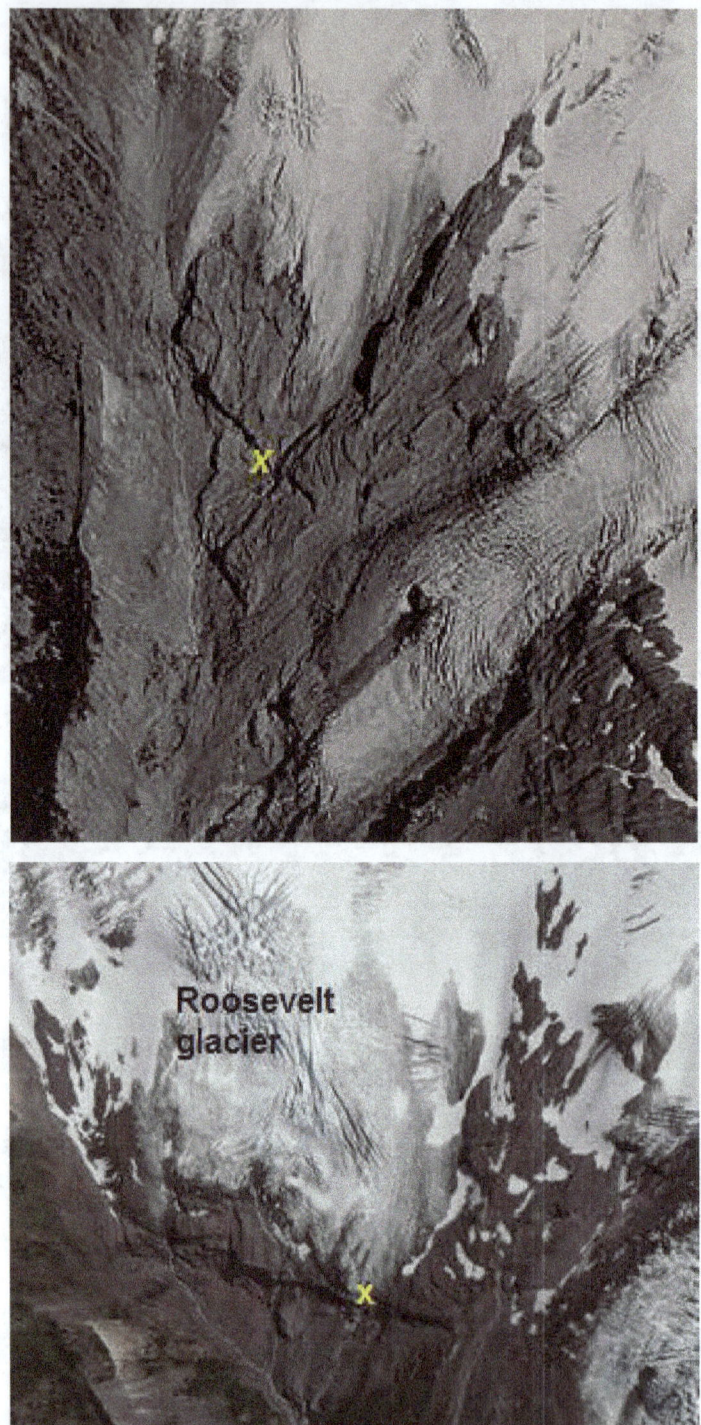

Figure 175. Comparison of photographs of the Roosevelt glacier in 1947 (top) and 2015 (bottom). Note that the terminus of the glacier in 2015 is well downvalley from it was in 1947, i.e., the glacier is more extensive now than it was in 1947. The yellow X is a common point of reference.

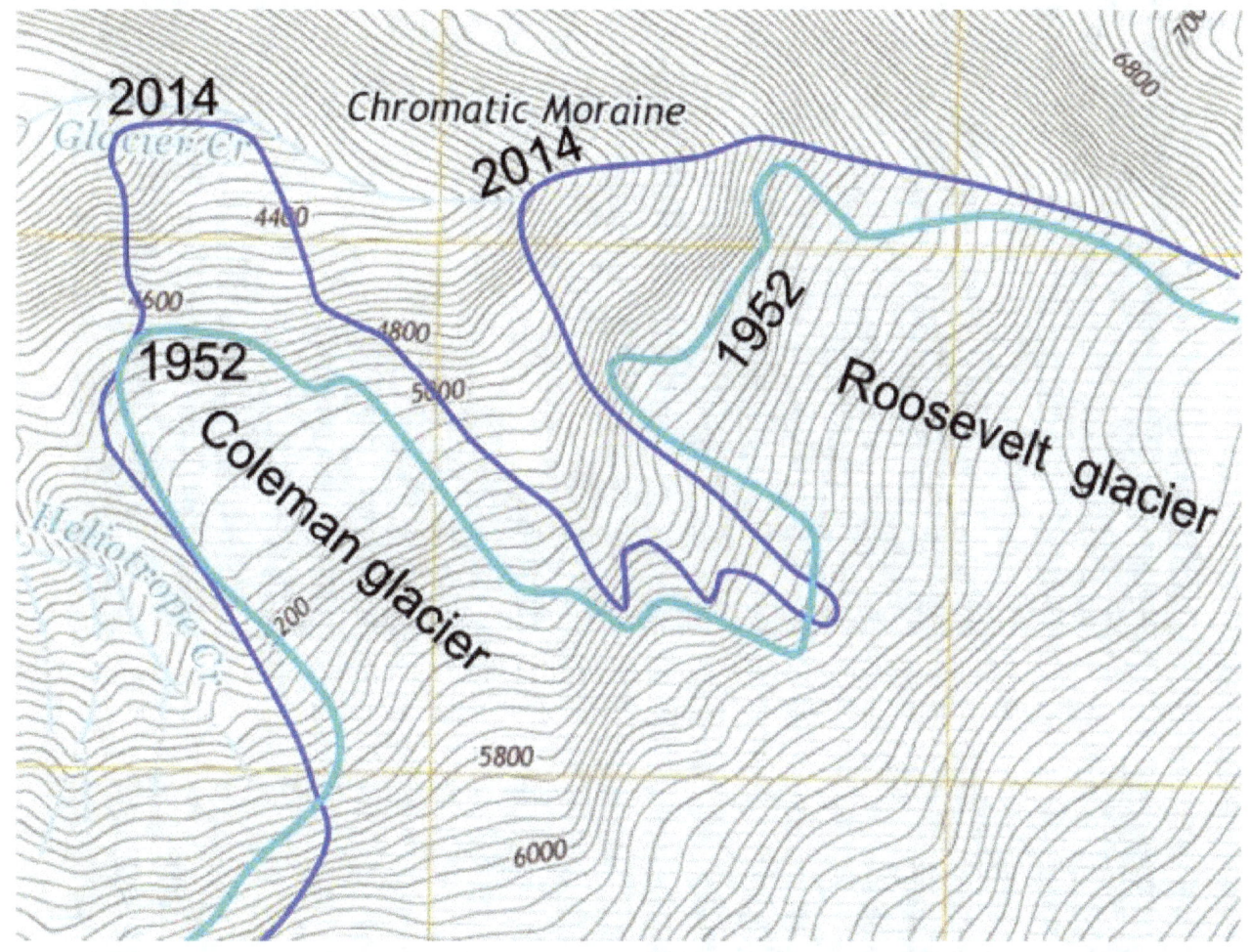

Figure 176. Comparison of terminal positions of the Coleman and Roosevelt glaciers 1952 (green line) and 2014 (blue line).

2011-2015 readvance of the Coleman and Roosevelt glaciers

Photos of the Roosevelt and Coleman glaciers on Mt. Baker, taken in 2011, 2013, and 2015 show that both glaciers had stopped retreating and advanced slightly from 2011 to 2015. Figures 177 and 178 show the position of the terminus of the Roosevelt glacier in 2011 and 2013. Note the bare ground at the red arrow in 2011 and the same area covered by ice in 2013 (Fig. 178).

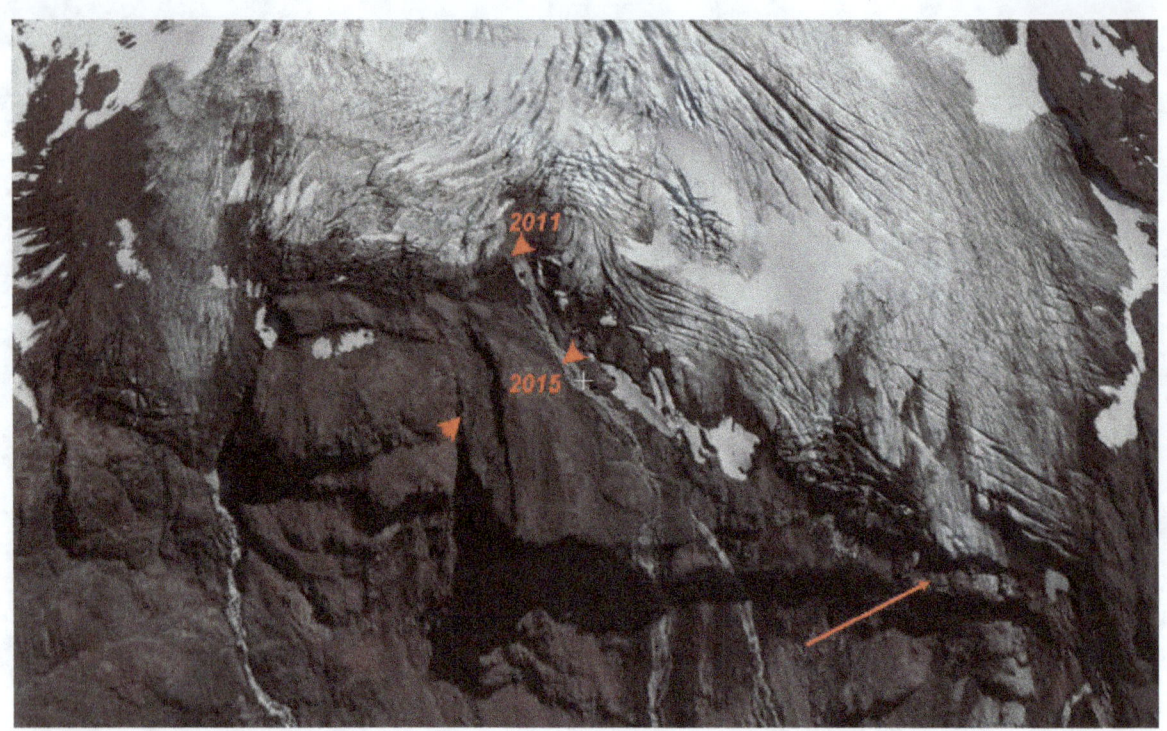

Figure 177. Terminus of the Roosevelt glacier in 2011. The red arrow points to bare ground beyond the glacier terminus. The red pointers indicate the glacier terminus in 2011 and 2015.

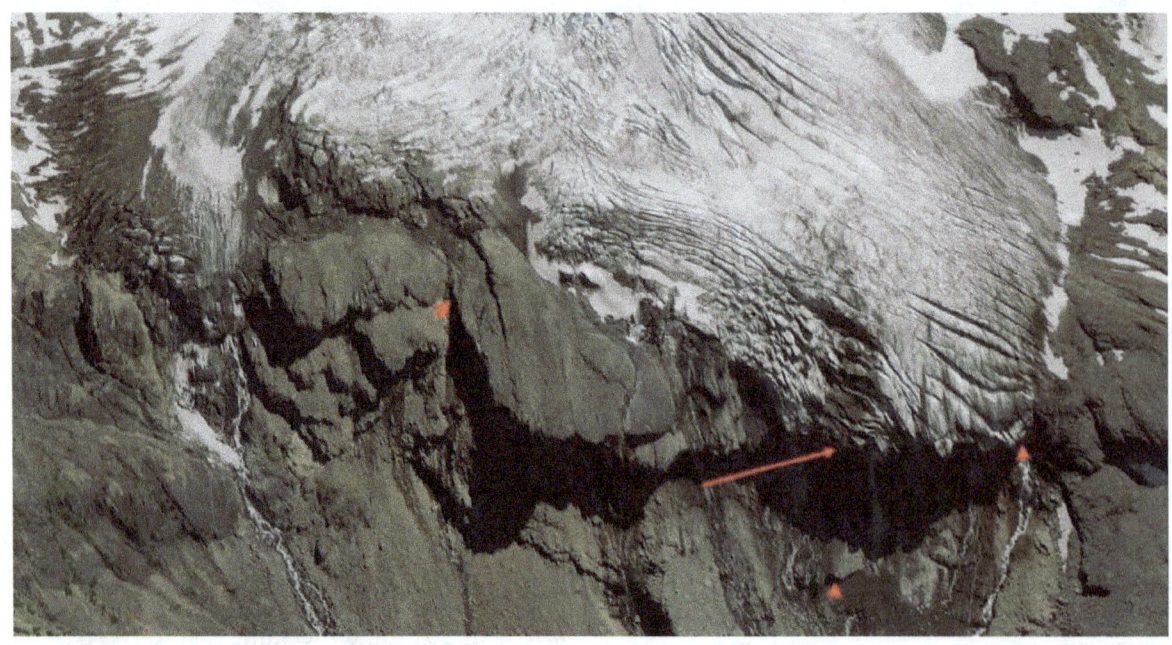

Figure 178. Terminus of the Roosevelt glacier in 2013. The red arrow points to ice now covering ground that was bare in 2011.

Figures 179 and 180 show the position of the terminus of the Coleman glacier in 2011 and 2013. Note the bare ground at the red circles in 2011 and the same area covered by ice in 2013. The lobe of ice at the left margin of the glacier moved even farther downvalley in 2015.

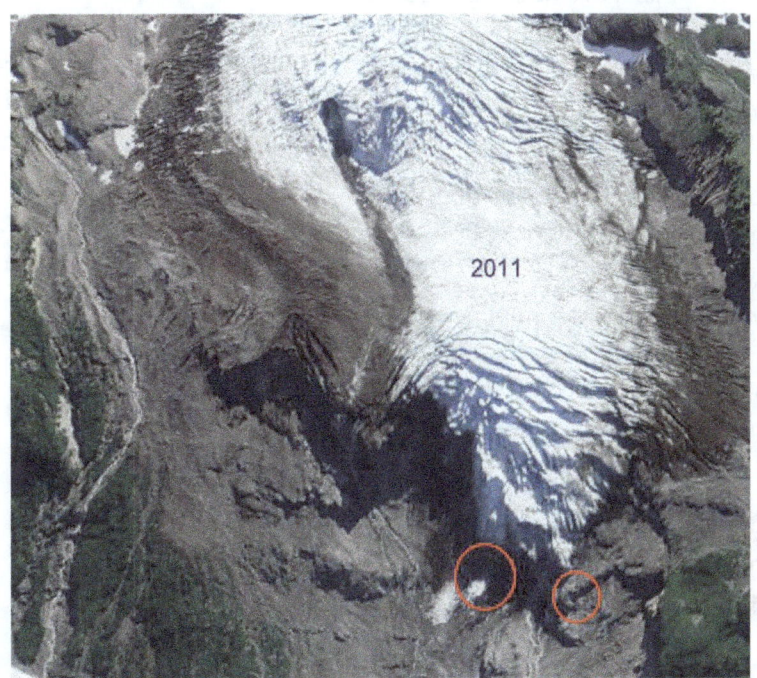

Figure 179. Terminus of the Coleman glacier in 2011.
Note the bare ground at the red circles.

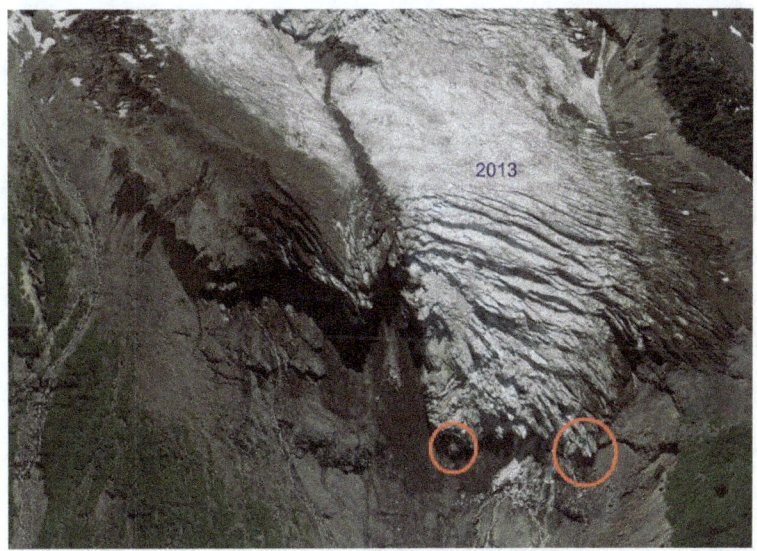

Figure 180. Terminus of the Coleman glacier in 2013.
Note that ice now covers bare ground shown in the 2011 photo.

Temperatures for the preceding decade

The cause of recent advance of the Roosevelt and Coleman glacier termini appears to be cooling in the western Cascades during the preceding decade. Temperatures recorded by NOAA in the western Cascades for the decade prior to the 2013 advance are shown in Fig. 181. The data show a cooling trend of –2.5°F per decade from 2003 to 2013. The advance of the Roosevelt and Coleman glaciers shown on the 2013 photos are also seen on 2015 photos, despite a fairly warm summer in 2014 and low snowpack in the winter of 2014-2015. Although the advance so far is quite small, it is significant because the advance has persisted for 2-3 years and shows a reversal of the recessional trend from the 1980s. The advance may or may not continue into the future—time will tell.

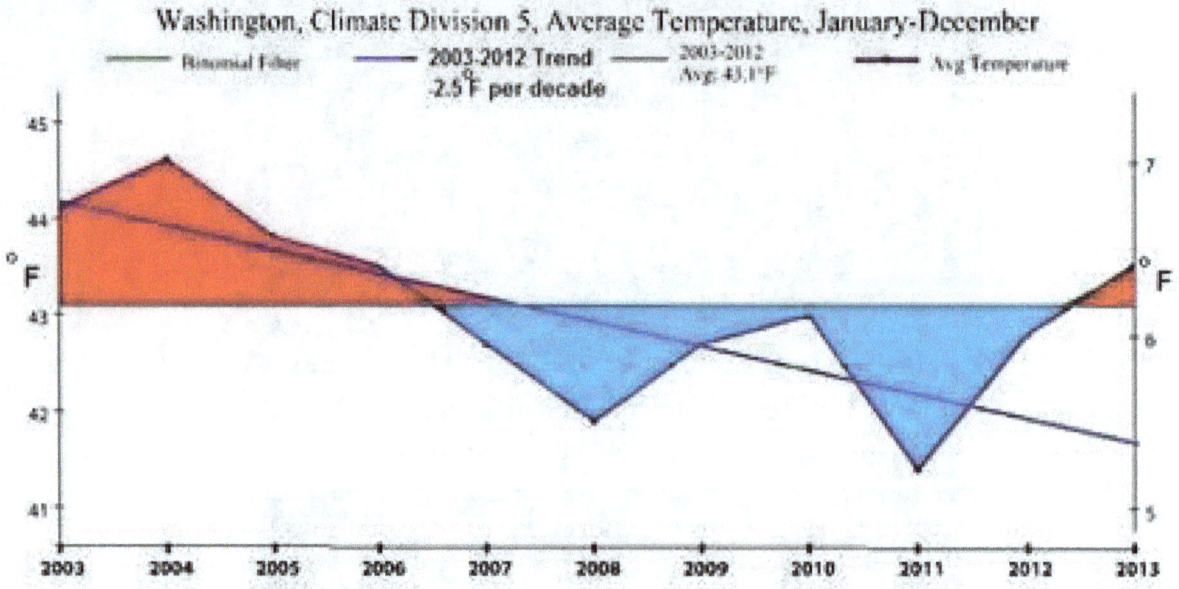

Figure 181. NOAA temperature measurements for the western Cascade Range for the decade prior to the 2013 advance of the Roosevelt and Coleman glaciers. The red part of the curve was warmer than the average annual temperature for the decade, the blue area was cooler. The straight line is the temperature trend for the decade, cooling of –2.5°F per decade.

Thunder Glacier

The Thunder glacier occupies a cirque cut into the Black Buttes volcanic cone on the western flank of Mt. Baker (Fig. 182). It is separated from the Coleman glacier by Heliotrope Ridge and from the Deming glacier by the ridge between Lincoln and Colfax Peaks.

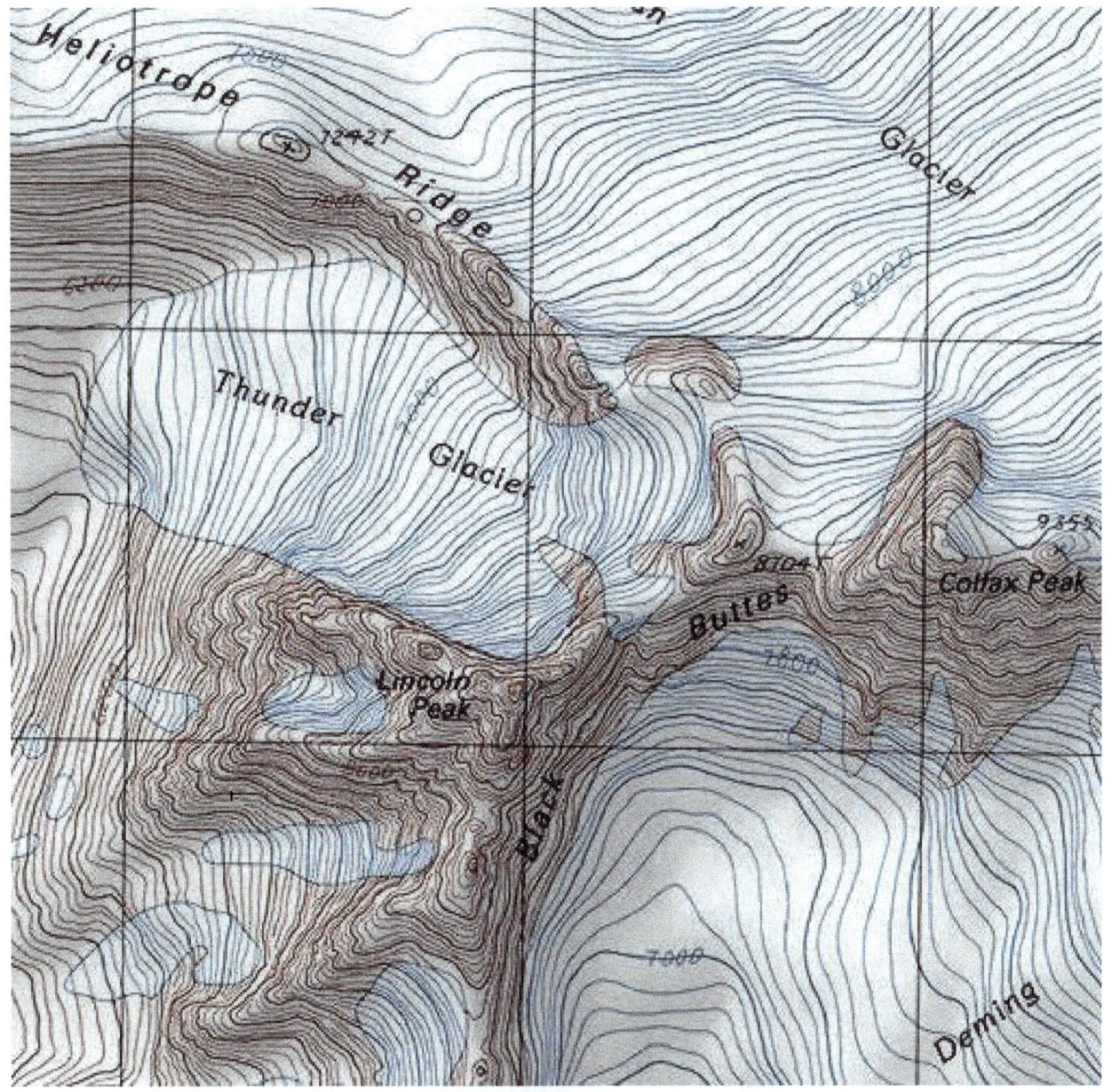

Figure 182. Thunder glacier.

Moraines downvalley from the present glacier are quite indistinct and even Little Ice Age terminus positions are not known. The earliest glacial terminus records are shown on the 1909 and 1915 USGS topographic maps (Figs. 183, 184) and 1940 air photographs (Fig. 185).

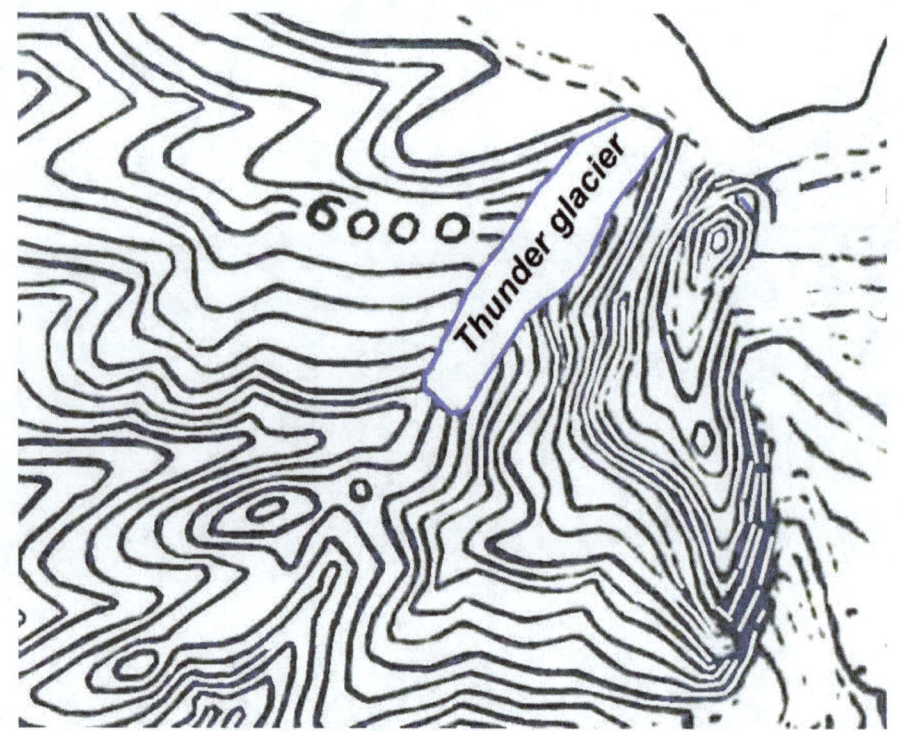

Figure 183. Thunder glacier, 1909. (USGS map)

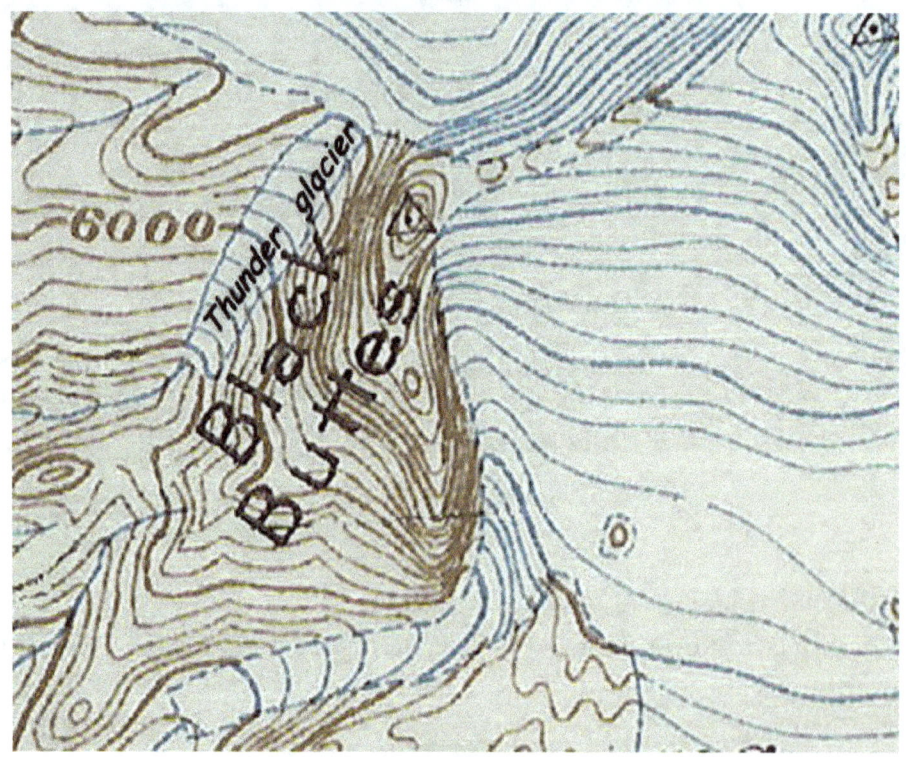

Figure 184. Thunder glacier, 1915.

Figure 185. Thunder glacier, 1940.

Figure 186. Thunder glacier, 1950.

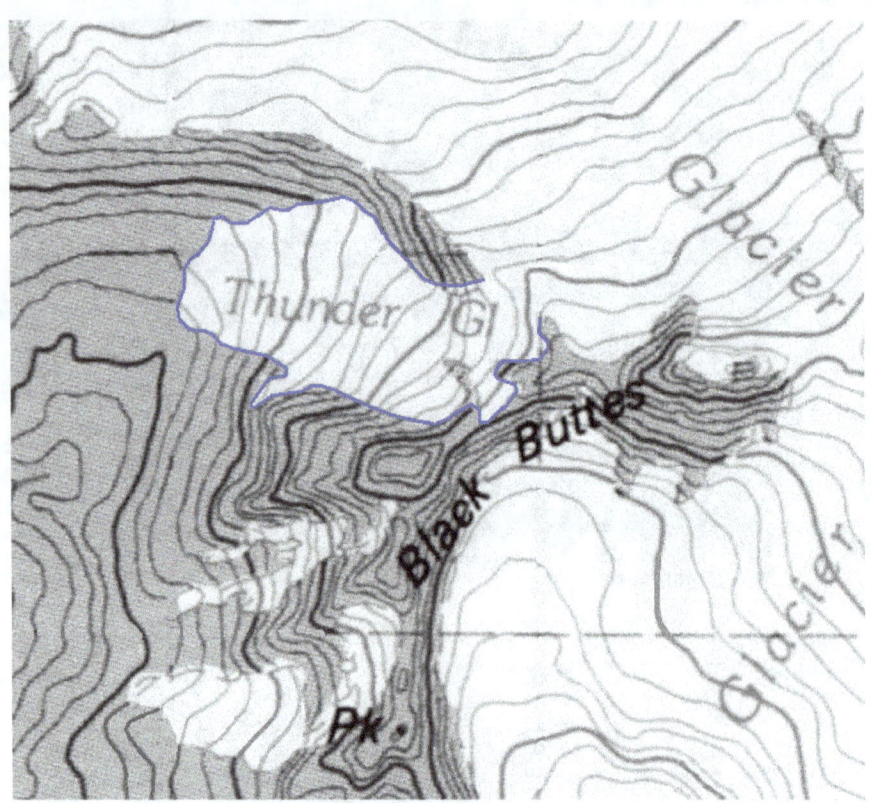

Figure 187. Thunder glacier, 1979.

Figure 188. Thunder glacier, 1987.

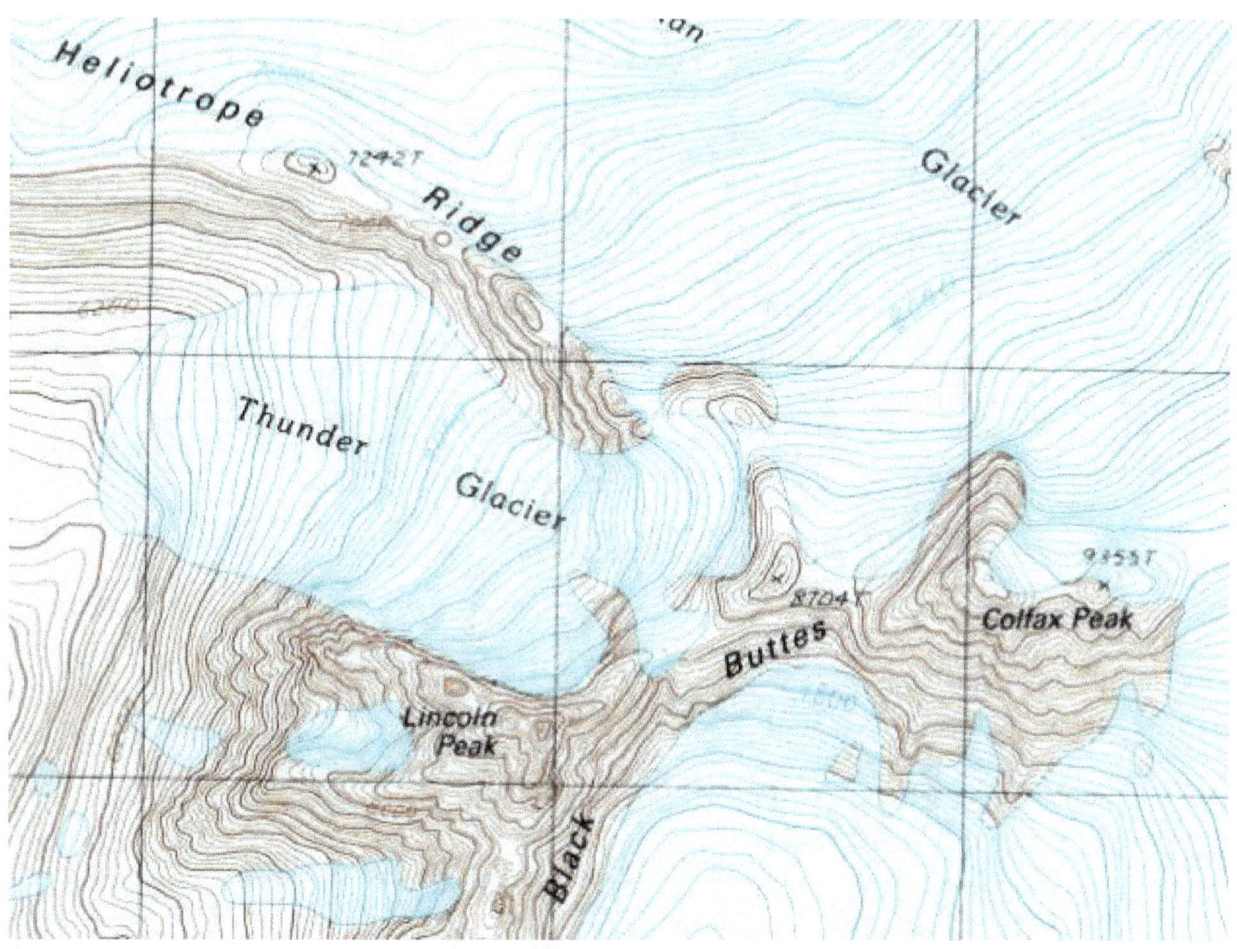

Figure 189. Thunder glacier, 1989.

Comparison of the 1952 terminal position with the 2014 position (Figs. 190, 191) shows that the Thunder glacier is now more extensive than it was in 1952.

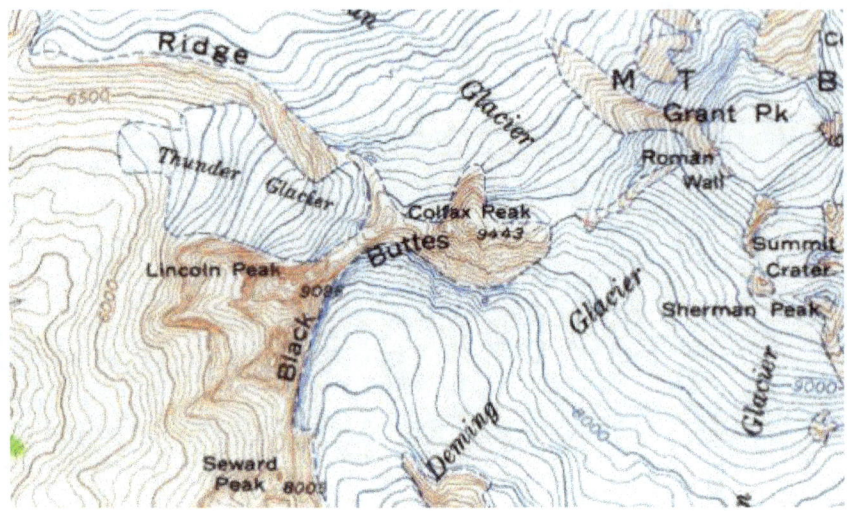

Figure 190. Thunder glacier, 1952.   (USGS map)

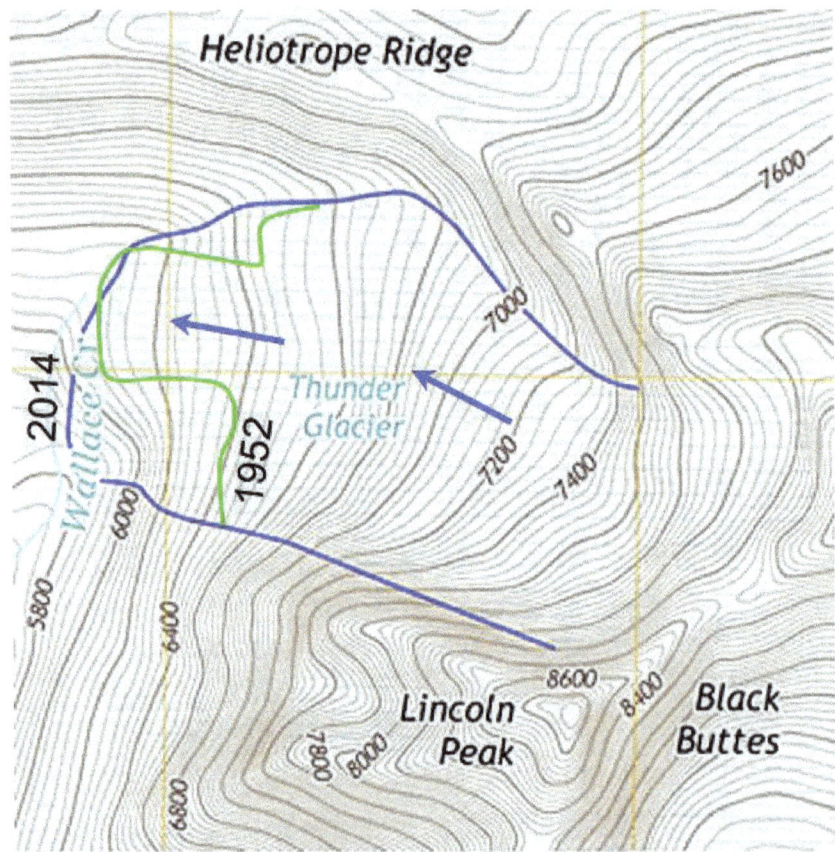

Figure 191. Comparison of 1952 (green line) and 2014 (blue line) termini. (USGS map)

Deming Glacier

The Deming glacier originates high on the main summit cone, flows into a deep cirque basin on the Black Buttes, then out of the cirque over a steep icefall into the lower valley (Figs. 192-196). Avalanches and rockfalls from the steep valley walls fall onto the glacier, so the sides of the lower glacier are usually covered with rock debris (Figs. 197-200). The Deming glacier is the source of the Nooksack Middle Fork, which is always muddy from the load of glacial silt that it carries. The lower glacier occupies a deep trough scoured out by ice.

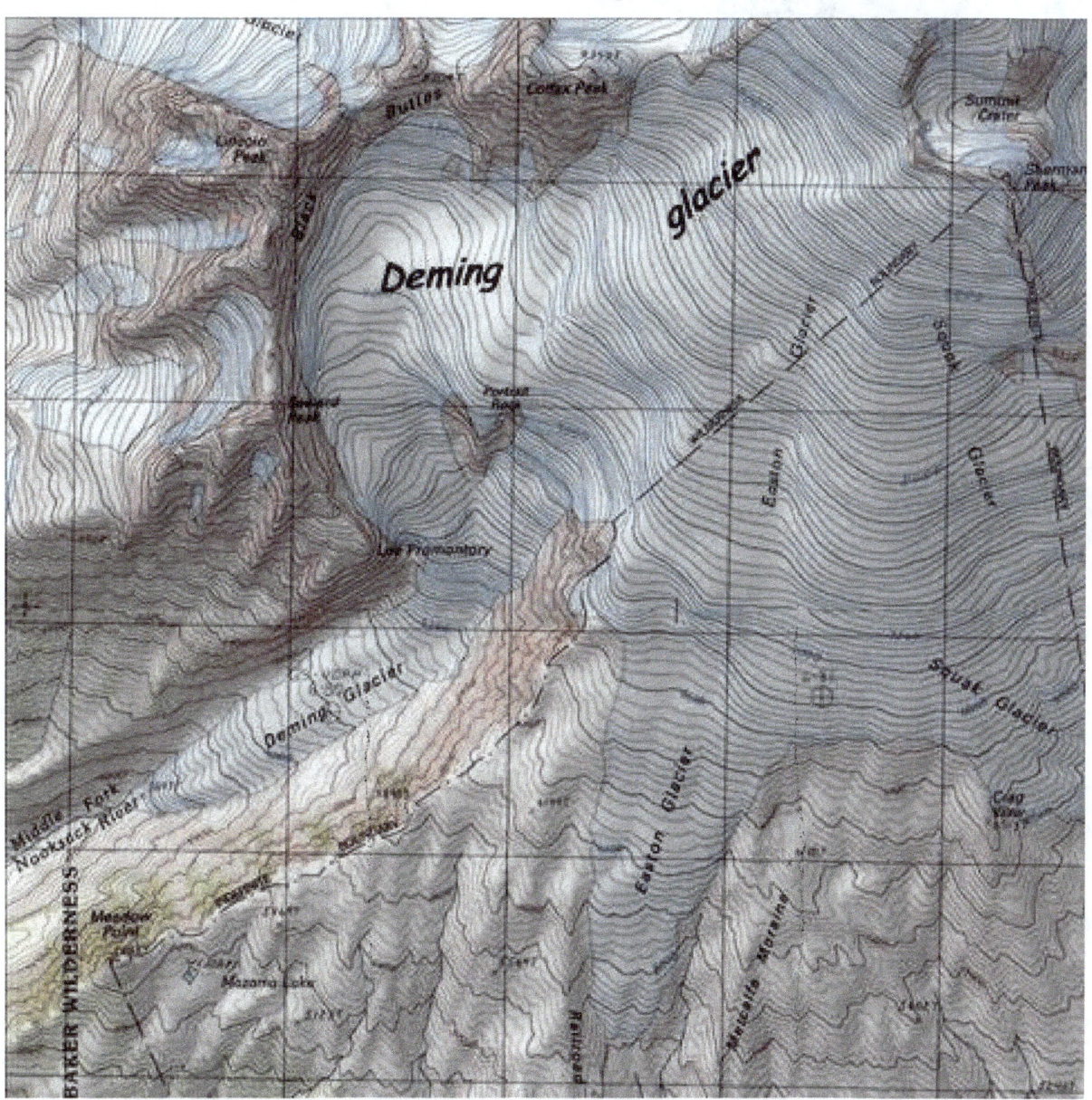

Figure 192. Topographic map of the Deming glacier

Figure 193. Deming glacier.

Figure 194. Deming glacier.

Figure 195. Deming glacier flowing from the summit cone into the Black Buttes cirque, then down a steep icefall to the valley floor.

Figure 196. Deming glacier icefall.

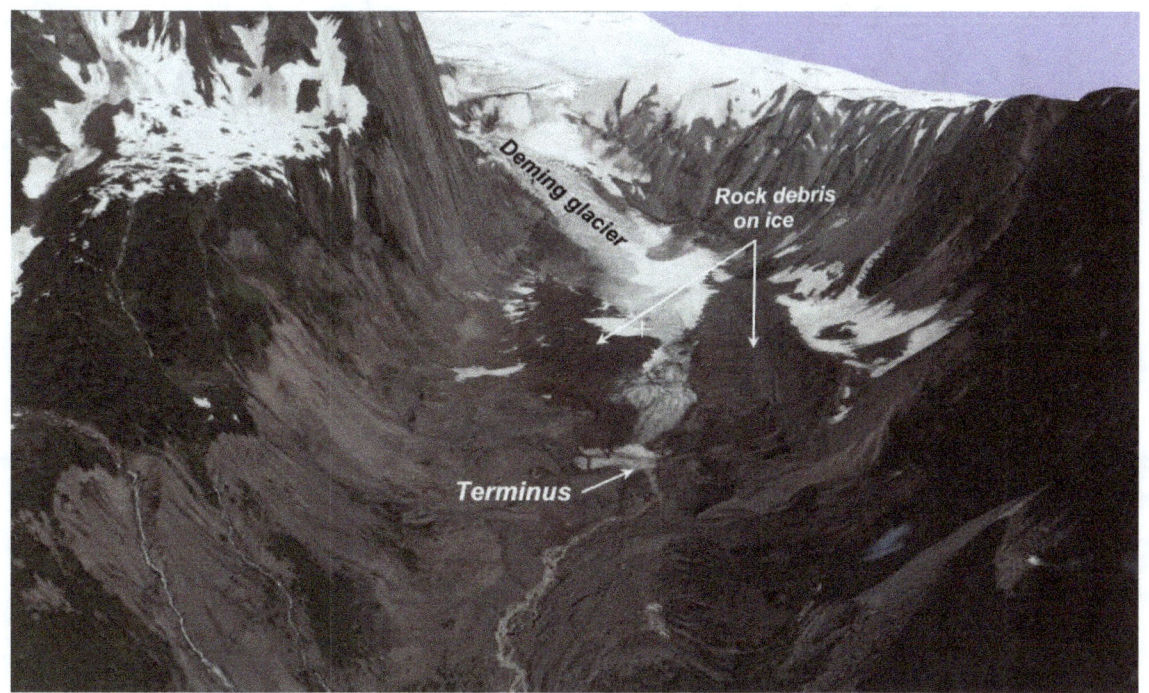

Figure 197. Lower Deming glacier.

Figure 198. Terminus of the Deming glacier in the 1970s.

Figure 199. Meltwater tunnel in the terminus of the Deming glacier in the 1970s. The terminus is quite steep because the glacier is close to the end of its 1979 maximum advance. Note the person at the bottom of the photo for scale.

Figure 200. Rock debris accumulating at the terminus of the Deming glacier, building an end moraine.

Late Ice Age moraines of the Deming glacier

Moraines in the Nooksack Middle Fork below the Deming glacier preserve a record the longest period of glacial advances on Mt. Baker. A high lateral moraine along the north side of the valley marks the margin of the glacier during the Younger Dryas cold period about 11–13,000 years ago (Figs. 201-210). Six logs buried in a lateral moraine along the south side of the lower end of the valley near Ridley Creek have been radiocarbon dated at ~10,500 ^{14}C years old (Fig. 203).

Logs and buried forests in lateral moraines occur farther upvalley at several localities. Logs at two localities between the present glacier terminus and Ridley Creek have been radiocarbon dated at 2,960 and 2,970 years old and 2,440 and 2,205 years old (Figs. 204-205). These buried logs imply that the Deming glacier extended at least that far downvalley at the time of the dates. Access to the sites of the buried logs is difficult because of lack of a trail, a deep gorge at the lower end of the valley, and growth of closely spaced alders.

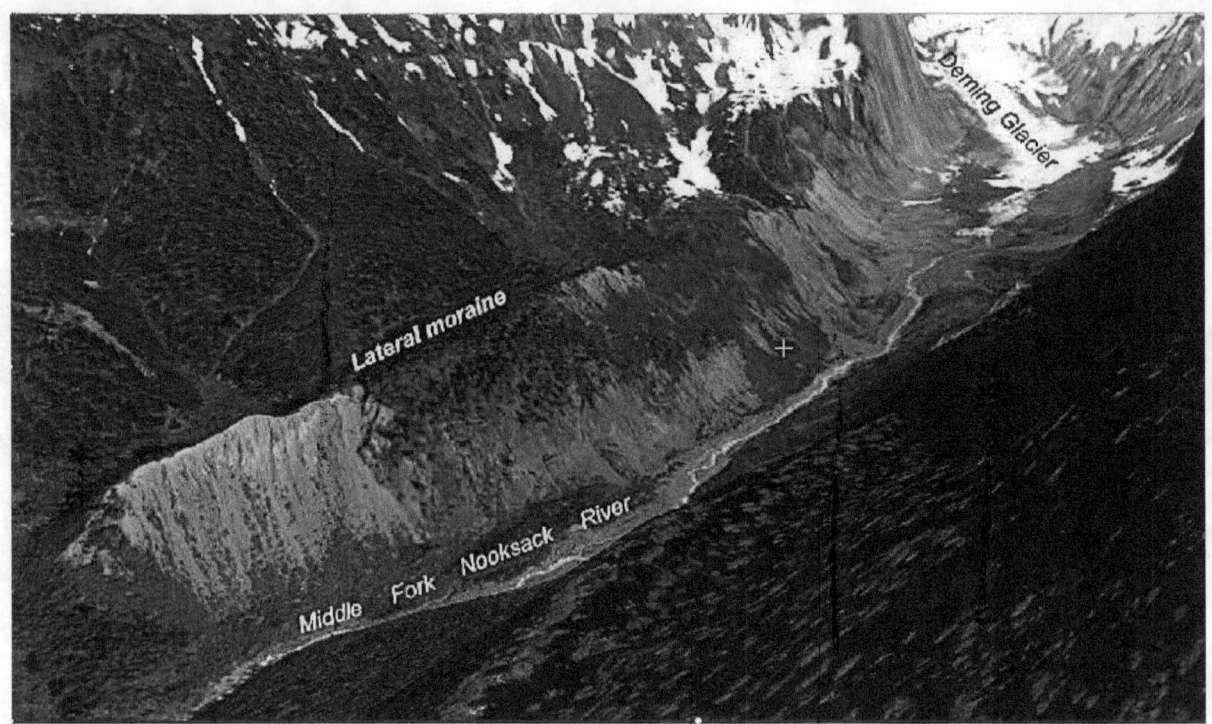

Figure 201. High lateral moraine along the north side of the Nooksack Middle Fork below the Deming glacier

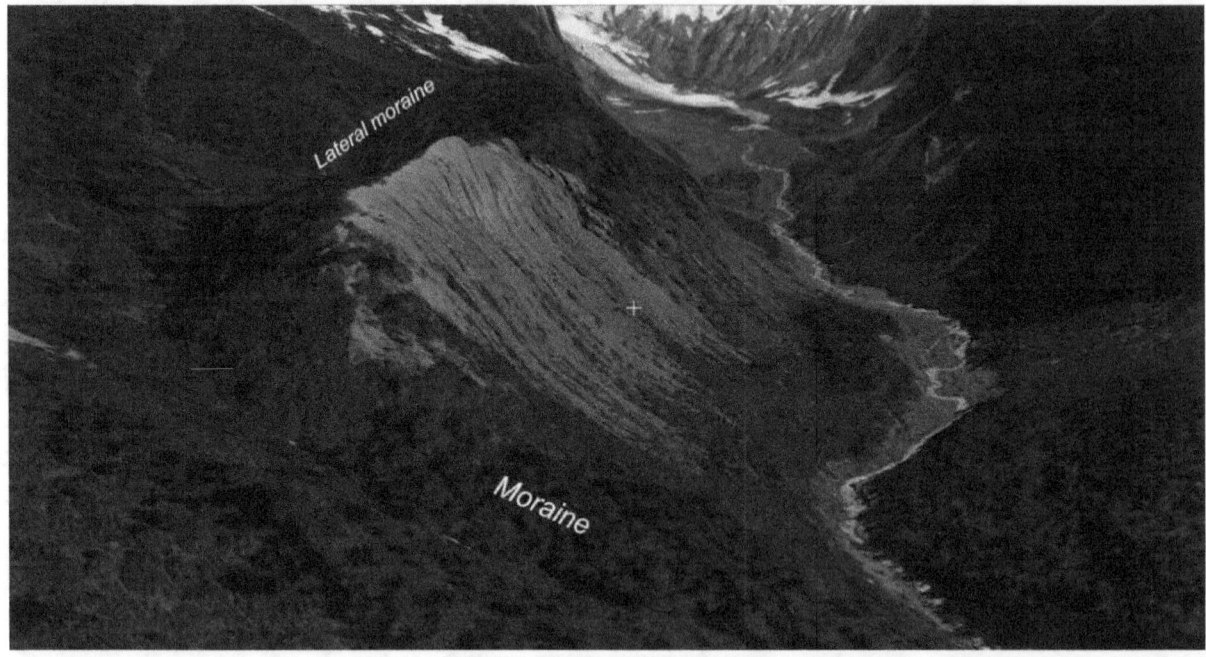

Figure 202. Little Ice Age lateral moraine below the Deming glacier. The lateral moraine bends around toward the axis of the valley, marking the former terminus of the glacier.

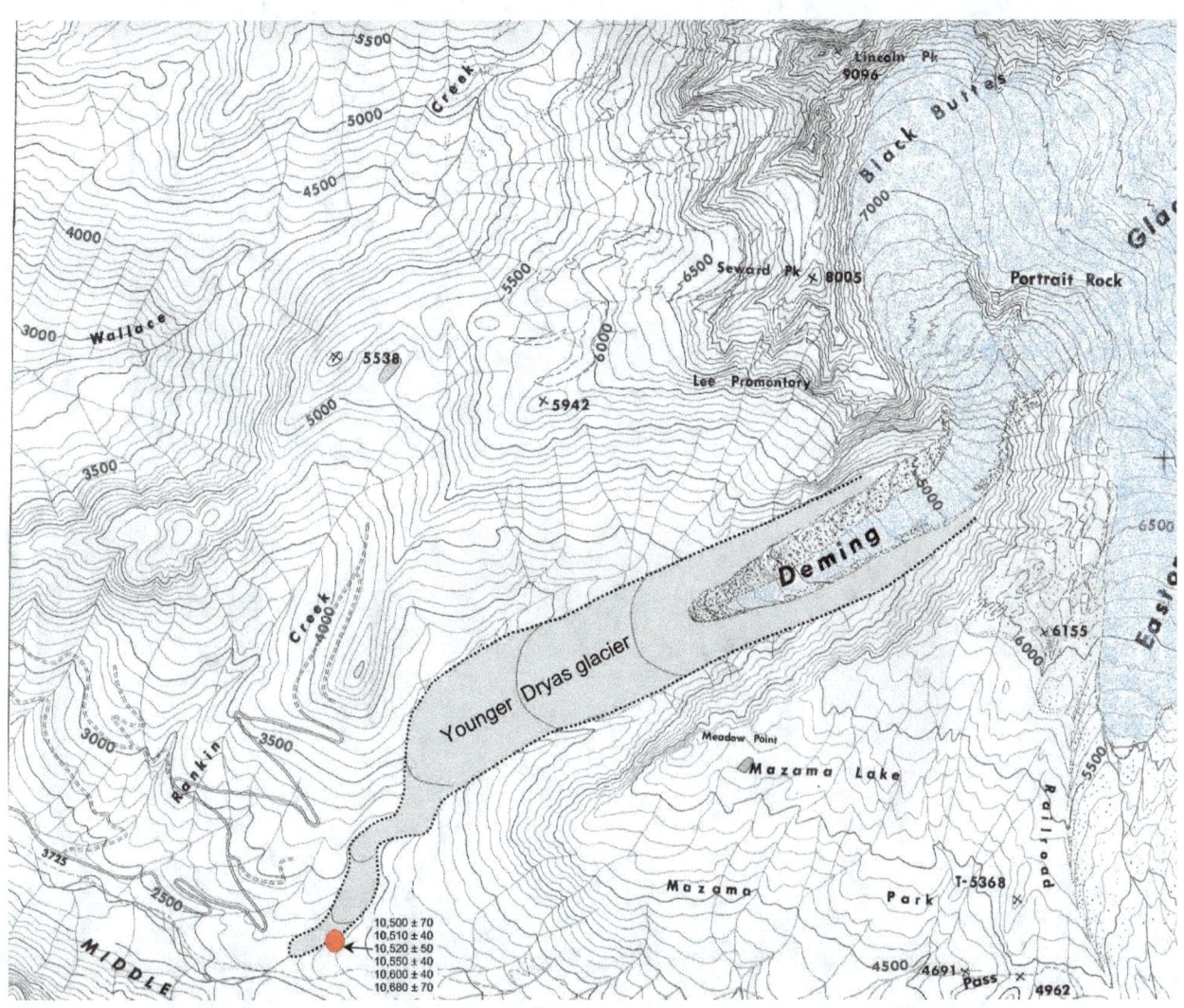

Figure 203. Reconstruction of the Deming glacier during the Younger Dryas cold period at the end of the last Ice Age about 11–13,000 years ago. Six radiocarbon dates from logs buried a moraine in the lower valley near Ridley Creek indicate an age of ~10,500 ^{14}C years. (Modified from USGS topographic map)

Figure 204. [top] Buried forest in a lateral moraine in the Nooksack Middle Fork above its junction with Ridley Creek. (Photo by Steve Fuller)

[bottom] Log buried in a lateral moraine in the same area. (Photo by Carrie Donnell)

Figure 205. Radiocarbon dates from logs buried in glacial deposits in the upper Middle Fork valley above Ridley Creek. (Carrie Donnell)

Little Ice Age Moraines of the Deming glacier

Isotope analyses of Greenland ice cores shows that for 8,500 years after the end of the last Ice Age, global climates were several degrees warmer than present. Then about 1,500 years ago, the climate cooled and initiated a cold period known as the Little Ice Age. During the Little Ice Age, alpine glaciers all over the world advanced downvalley and human populations suffered greatly from repeated crop failures and starvation.

The Deming glacier during the Little Ice Age about 500 years ago was not as extensive downvalley as during the Younger Dryas cold period of 11,000 years ago, but the ice reached the lateral moraine high above the valley floor (Figs. 206-210). Nested inside the highest lateral moraine are half a dozen younger moraines (Figs. 209, 210).

Figure 206. Lateral moraine high above the valley floor of the Deming glacier.

Figure 207. High lateral moraine and 1979 forest trim line in the lower Deming valley.

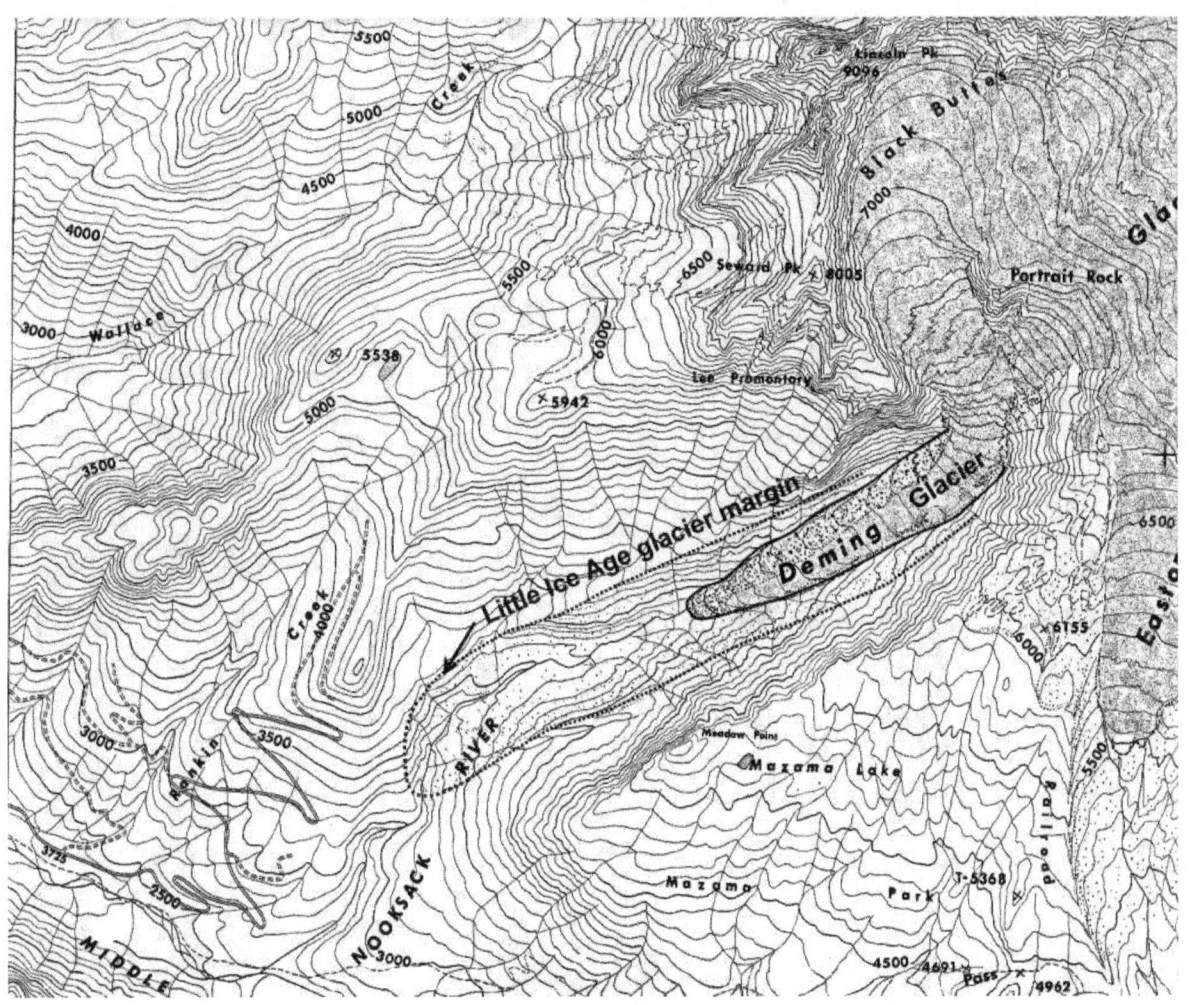

Figure 208. Former ice margin of the Deming glacier during the Little Ice Age. (Modified from USGS topographic map)

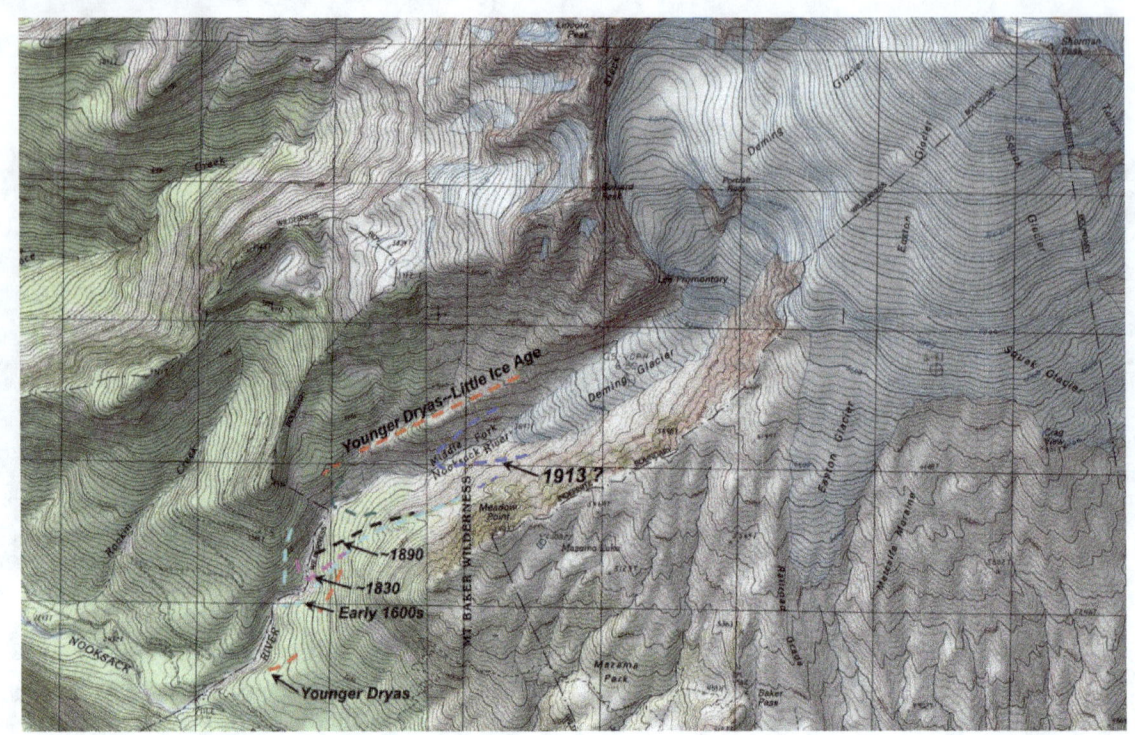

Figure 209. Former termini of the Deming glacier. (Modified from USGS topographic map)

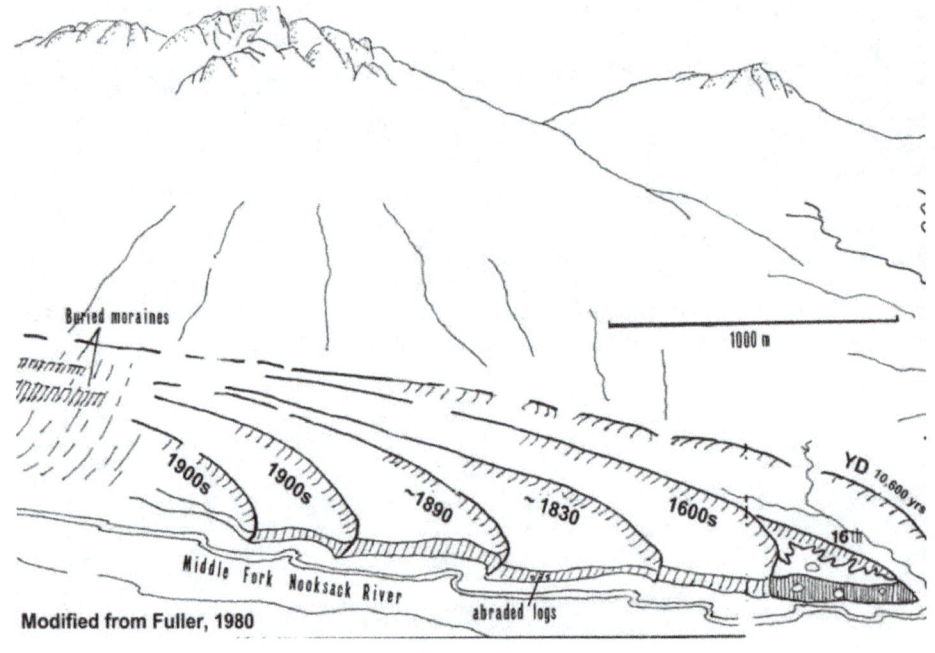

Figure 210. Moraines downvalley from the Deming glacier.

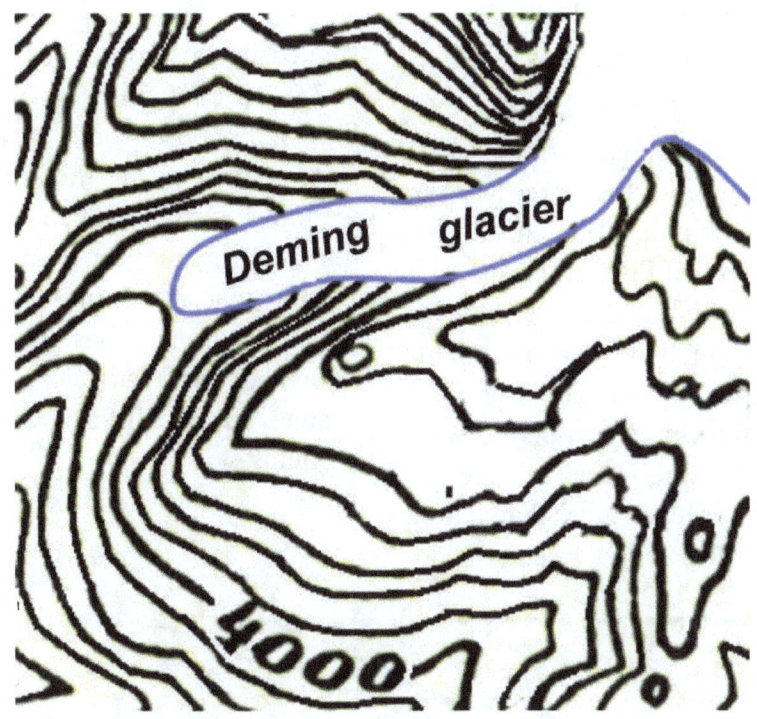

Figure 211. Deming glacier in 1909 ~3000 feet downvalley from the present terminus. (USGS topographic map).

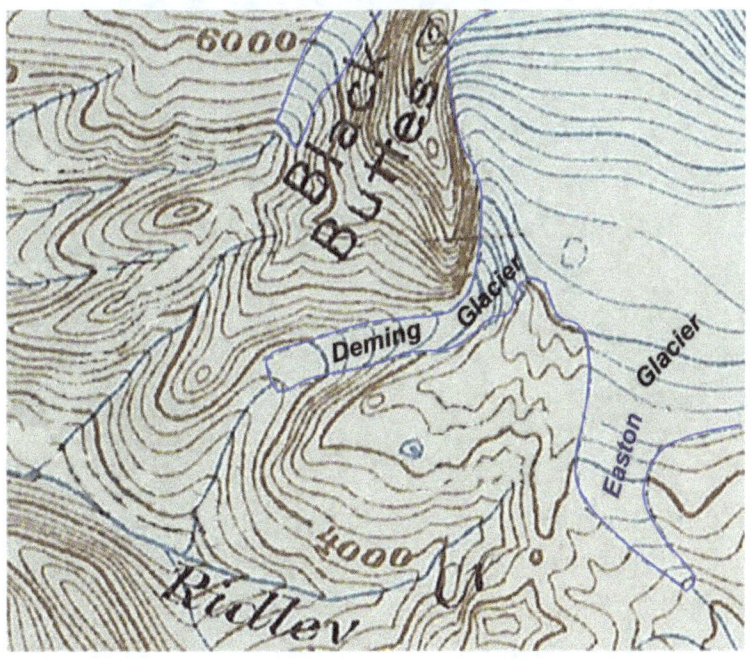

Figure 212. Deming glacier, 1915. (USGS topographic map).

Deming glacier fluctuations, 1909 to 2015

During the 1880 to 1915 cool period, the Deming glacier readvanced approximately 3000 feet downvalley from the present terminus, just upvalley from the position of the Little Ice Age terminus (Fig. 209). Figure 211 shows the Deming glacier in 1909 and Figure 212 shows it in 1915. Like the other Mt. Baker glaciers, the ice margin marks the position of the glacier terminus at the culmination of the global cool period from 1880 to 1915. The following warm period from 1915 to 1945 was the warmest period of the last century and the glacier receded more than a mile upvalley. About 1945, the Pacific Ocean offshore flipped into its cool mode and 30 years (1945–1977) of cool climate and glacier readvance followed. In 1977, the Pacific abruptly flipped from cool to warm, initiating another period of warming during which the Deming glacier retreated upvalley once again. So although the Deming glacier terminus is now upvalley from its earlier positions, it has oscillated back and forth, retreating and readvancing with every climate change, but is still far downvalley from its 1952 terminus.

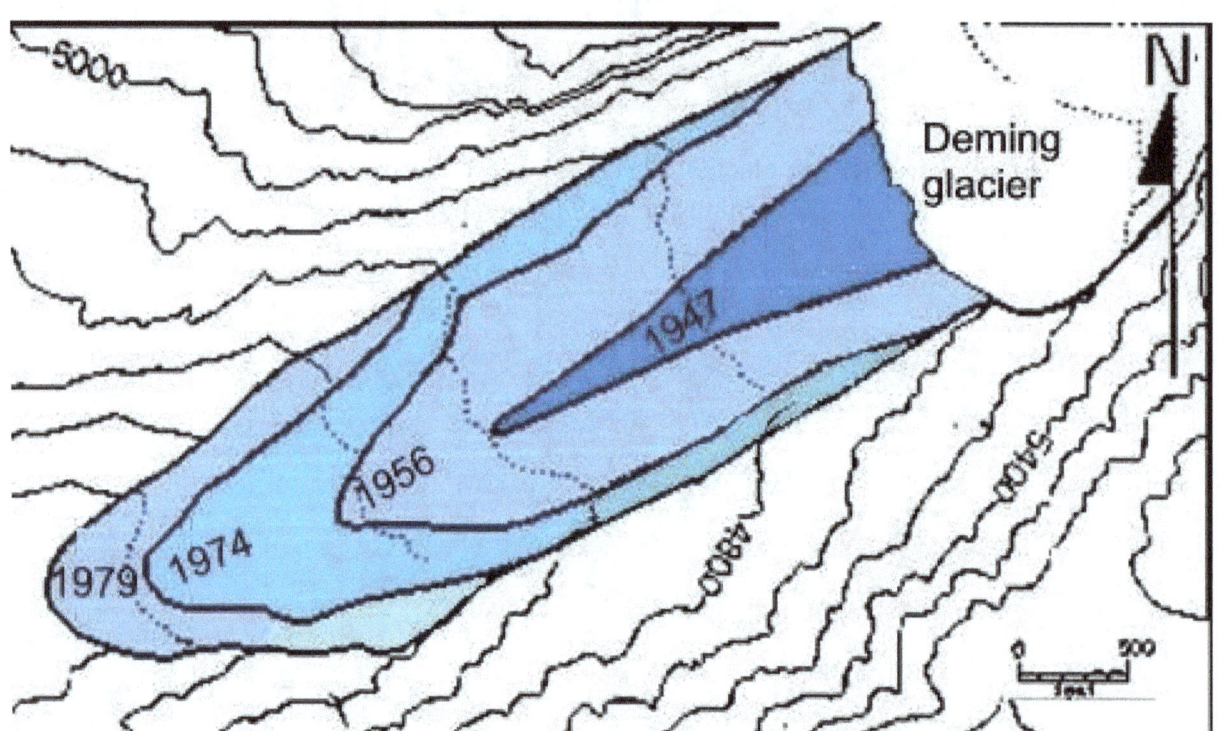

Figure 213. Retreat of the Deming glacier from 1940 to 1947 near the end of the ~1915 to ~1945 warm period. The Deming terminus retreated 1300 feet from 1940 to 1947, then began to readvance. (Modified from Harper, 1992)

The terminus of the Deming retreated rapidly upvalley during the 1915 to 1945 warm period. The most rapid retreat probably occurred during the 1930s, the warmest decade of the century, but photographs of that time are very rare. The earliest photo records begin in 1940 so only the last part of retreat record is known. From 1940 to 1947 the Deming glacier receded 1296 feet (Figs. 213- 215). The retreat ended by 1947 when the climate cooled and the glacier began to readvance. From 1947 to 1963, the terminus advanced 1320 feet (Fig. 213). From 1963 to 1970, the glacier advanced 265 feet and from 1970 to 1979 advanced 475 feet, culminating the 1947 to 1979 advance phase and initiating a new retreat phase. From 1979 to 1987, the Deming glacier retreated 470 feet and 144 feet from 1987 to 1990. Since 1990, the glacier has continued to retreat upvalley (Figs. 221-224).

Figure 214. Deming glacier in 1940.

Figure 215. Deming glacier in 1947. Note that the terminus continued to retreat upvalley from its 1940 position.

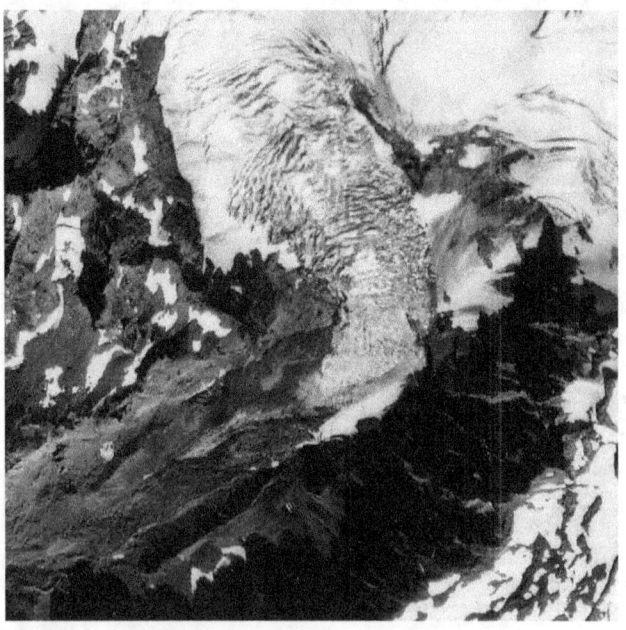

Figure 216. Deming glacier in 1950. The terminus has retreated slightly from its 1947 position and the glacier shows signs of stagnation in its lower part.

Figure 217. Deming glacier, 1962. The glacier terminus advanced 1320 feet to this position since 1947. (Photo by Austin Post)

Figure 218. Deming glacier, 1972. The terminus has continued to advance. (USGS photo)

Figure 219. Deming glacier at its maximum advance, 1979. (Photo by Austin Post)

Figure 220. 1979 terminus of the Deming glacier. Note the vegetation line just beyond the terminus. This line marks the position of the 1979 ice in more recent photos

Figure 221. Deming glacier, 1991. Note that it has retreated upvalley from the forest trim line that marks its 1979 terminus. (Photo by Austin Post)

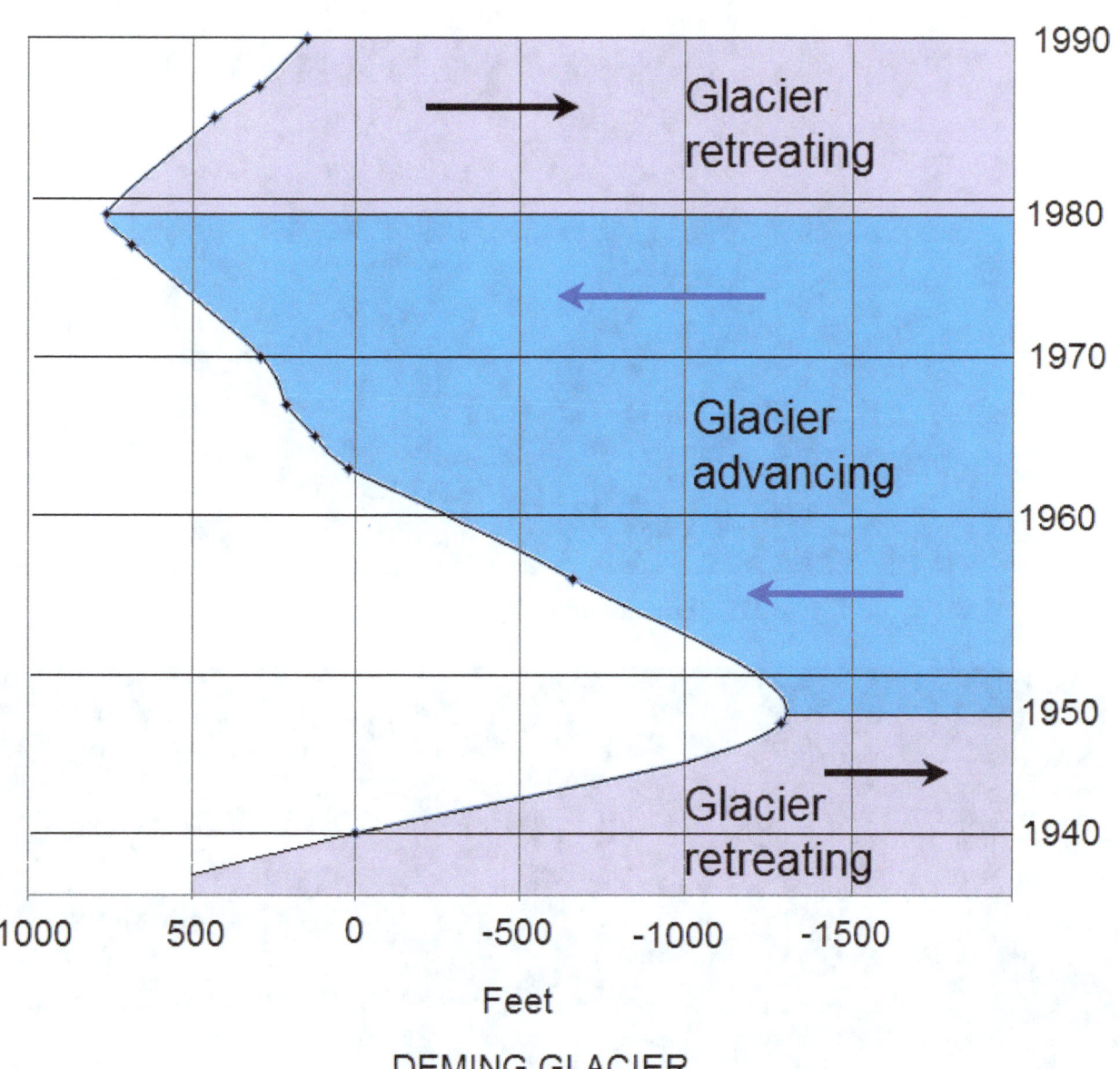

Figure 222. Graph of advance and retreat phases of the Deming glacier.

Figure 223. Terminus of the Deming glacier, 2009.

Figure 224. Terminus of the Deming glacier, 2015.

Figure 225 shows the position of the Deming glacier terminus in 1950 and in 2011. These photos indicate that the Deming glacier is more extensive now than it was in 1950. Figure 226 shows a comparison of the positions of the 1950 and 2014 USGS and 2014 glacial termini taken from 1952 topographic maps. The 2014 terminus is more than half a mile downvalley from its 1952 position (Fig. 226).

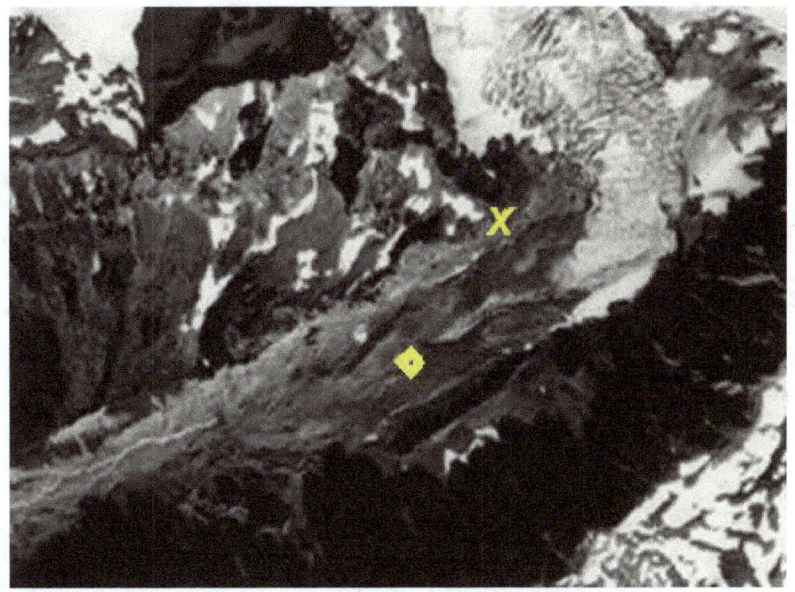

Figure 225. Deming glacier in 1950. The yellow X is a common point of reference and the diamond shape is the position of the terminus.

Figure 226. Deming glacier in 2011. Note how much farther downvalley the 2011 terminus is than the 1950 terminus

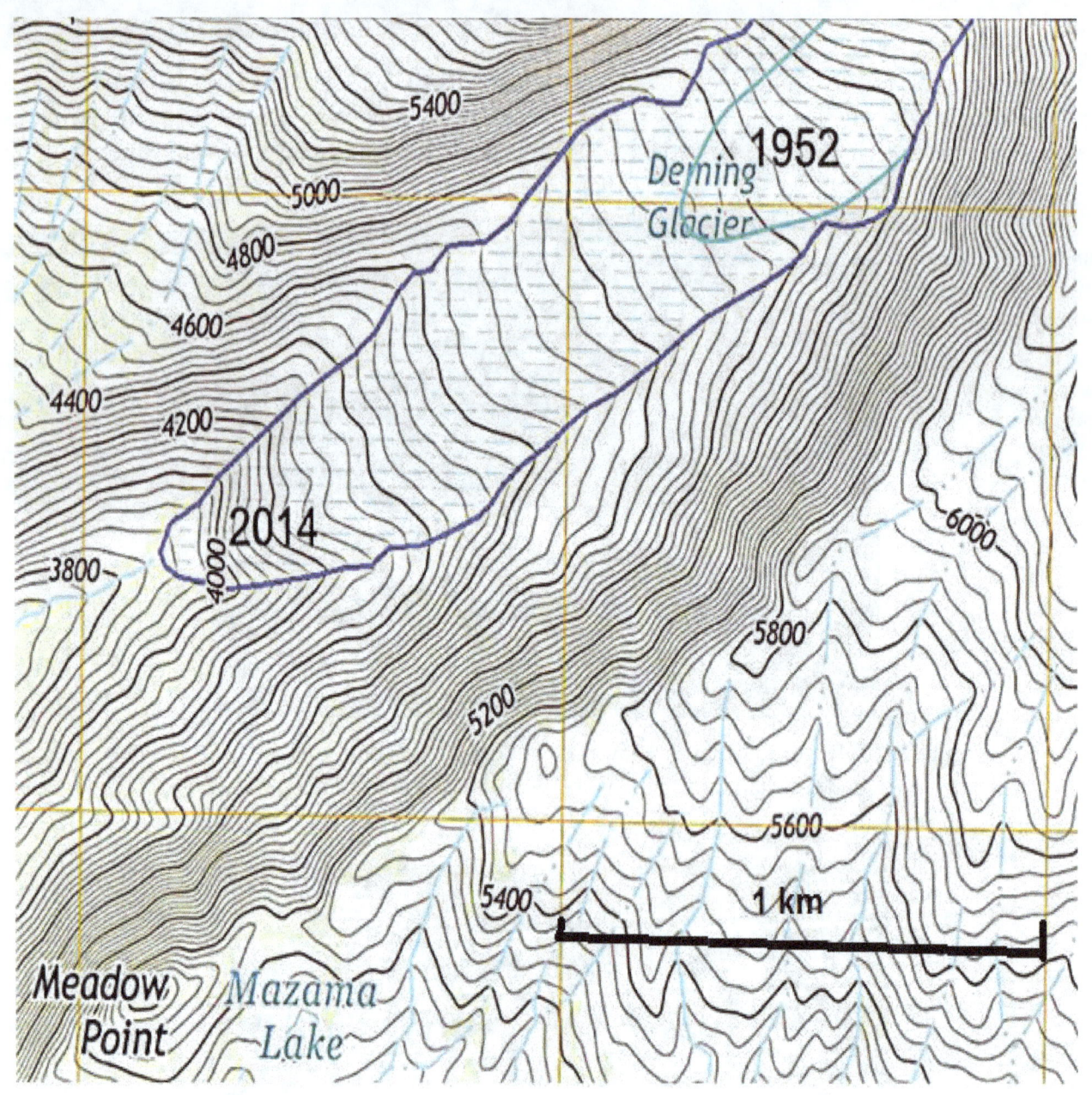

Figure 227. Comparison of the terminal position of the Deming glacier in 1952 (green line) and 2014 (blue line). The glacier in 2014 was ¾ of a mile farther downvalley than its 1952 position.

2013

Figure 228 shows the 2013 terminus of the Deming glacier at the top of a ledge that crosses the valley floor. The Deming glacier has stopped retreating and may even have advanced slightly, although the exact position of the terminus at the end of the summer melt season is difficult to see.

Figure 228. Deming glacier, 2013.

Easton Glacier

The Easton glacier is named after Charles Easton, an early pioneer. The glacier originates high on the south flank of Mt. Baker below Sherman crater and extends downvalley to the headwaters of Sulphur Creek (Fig. 229). High lateral moraines extend about a mile downvalley from the present glacier (Figs. 230-235). The western lateral moraine, known as Railroad Grade, has an exceptionally scenic trail along the crest (Fig. 235).

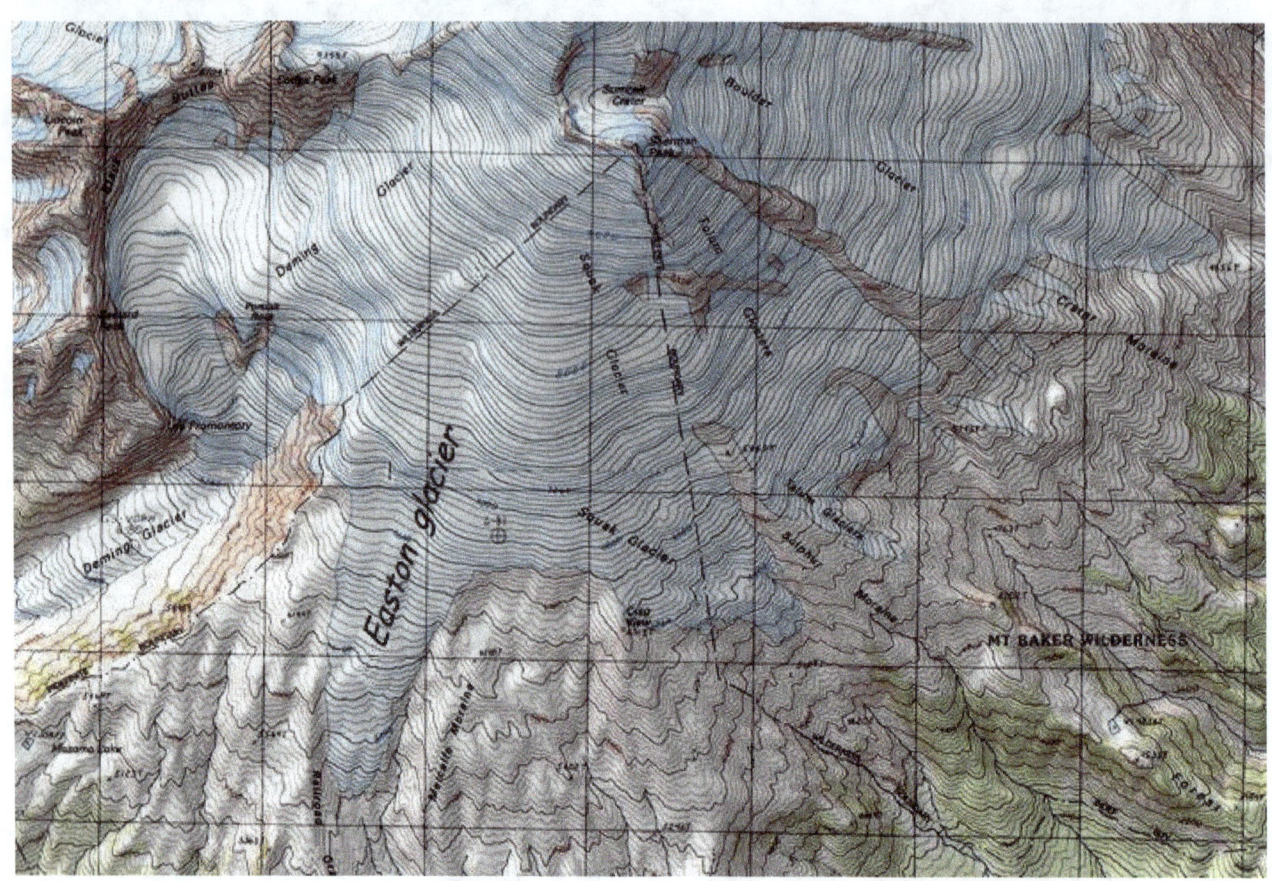

Figure 229. Topographic map of the Easton glacier. (USGS topographic map)

Figure 230. Easton glacier. (Photo by Austin Post)

Figure 231. Easton glacier and lateral moraines looking downvalley.

Figure 232. Railroad Grade lateral moraine viewed from the meadow to the west. The peak in the background is Lincoln Peak of the Black Buttes.

Figure 233. Brilliant fall vegetation on Railroad Grade, Easton glacier lateral moraine.

Late Ice Age moraines of the Easton glacier

Near the end of the last Ice Age about 15,000 years ago, temperatures abruptly warmed as much as 20°F in less than 100 years and the gigantic ice sheets covering much of the Northern Hemisphere landmasses began drastically melting. Then, 11–12,500 years ago, the climate suddenly cooled again (known as the Younger Dryas cold period) and the much–reduced glaciers readvanced, but not nearly to their former size. At Mt. Baker, the Cordilleran Ice Sheet had largely melted away during the sudden warming 15,000 years ago. During the Younger Dryas cold period, the terminus of the Easton glacier extended below timberline just above Schreibers Meadow and built large lateral moraines (e.g., Railroad Grade, Figs. 231–236). About 11,000 years ago, temperatures again soared 20°F in less than 100 years, ending the Ice Age. During the following 10,000 years, temperatures were warmer than at present and glaciers on Mt. Baker melted far upvalley.

Figure 234. Lateral moraine (Railroad Grade) of the Easton glacier.

Figure 235. Railroad Grade lateral moraine of the Easton glacier.

Figure 236. Railroad Grade, a lateral moraine of the Easton glacier.

Little Ice Age moraines

After about 8,500 years of temperatures higher than present, the climate began alternately warming and cooling about 1.500 years ago. Temperatures cooled between 500 AD and about 900 AD during a harsh period known as the Dark Ages. From about 900 AD to 1300 A, the climate warmed again (the Medieval Warm Period) and temperatures were slightly warmer than present. About 1300 AD, the climatic suddenly turned cold again and the Little Ice Age began. These changes are recorded in the glacial record of some of Mt. Baker's glaciers, especially the Coleman glacier where a buried forest about 700 years old represents the Medieval Warm Period between two colder periods of moraine building.

During the Little Ice Age (1300 to 1915 AD), the Easton glacier probably extended about a mile downvalley from its present terminus (Fig. 237) where annual rings from a tree on a small moraine indicate an age early in the century. Whether or not the ice reached the crest of the Railroad Grade lateral moraine is uncertain.

Glacial fluctuations between 1909 and 2015

In 1909, the terminus of the Easton glacier was about a mile downvalley from the present terminus (Figs. 237-240) as a result of the global cool period from 1880 to 1915. This terminal position of the glacier was close to the maximum extent of the glacier during the Little Ice Age about 500 years ago.

The global climate warmed from 1915 to 1945 and the Easton glacier retreated about a mile and a half upvalley. (Figs. 238–241). From 1934 to 1937, the warmest years of the 20th century, the Easton glacier retreated 475 feet (Fig. 242). From 1937 to 1940, the glacier receded 430 feet and from 1940 to 1947 it retreated 453 feet (Figs. 243, 244). Most of the other Mt. Baker glaciers had stopped receding and begun to advance by 1947, but the East continued to retreat into the early 1950s. It retreated 550 feet from 1947 to 1952 and 807 feet from 1947 to 1956 (Fig. 245).

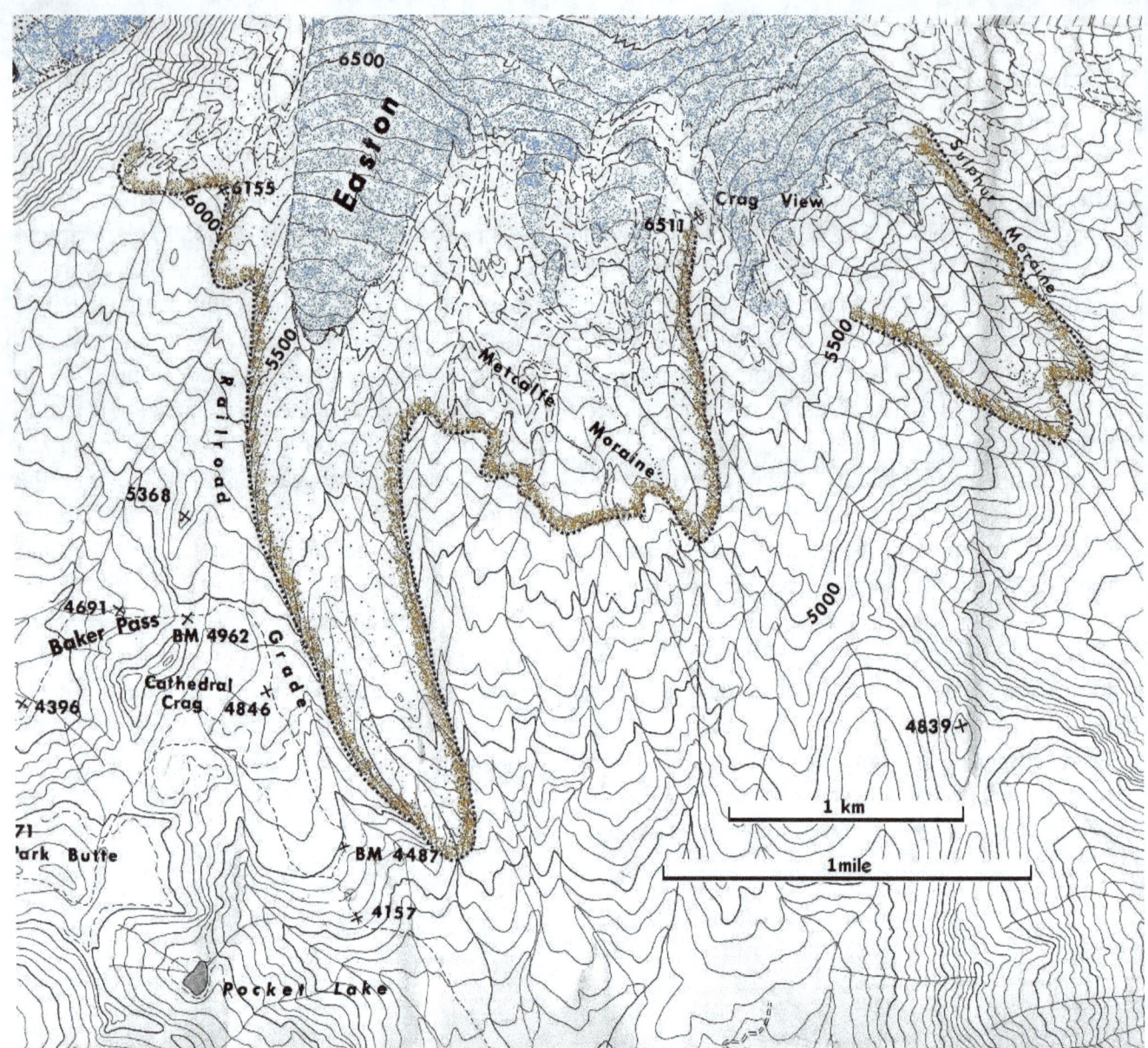

Figure 237. Little Ice Age (~1500 AD) glacier margins (brown pattern) of the Easton glacier. (Modified from USGS map)

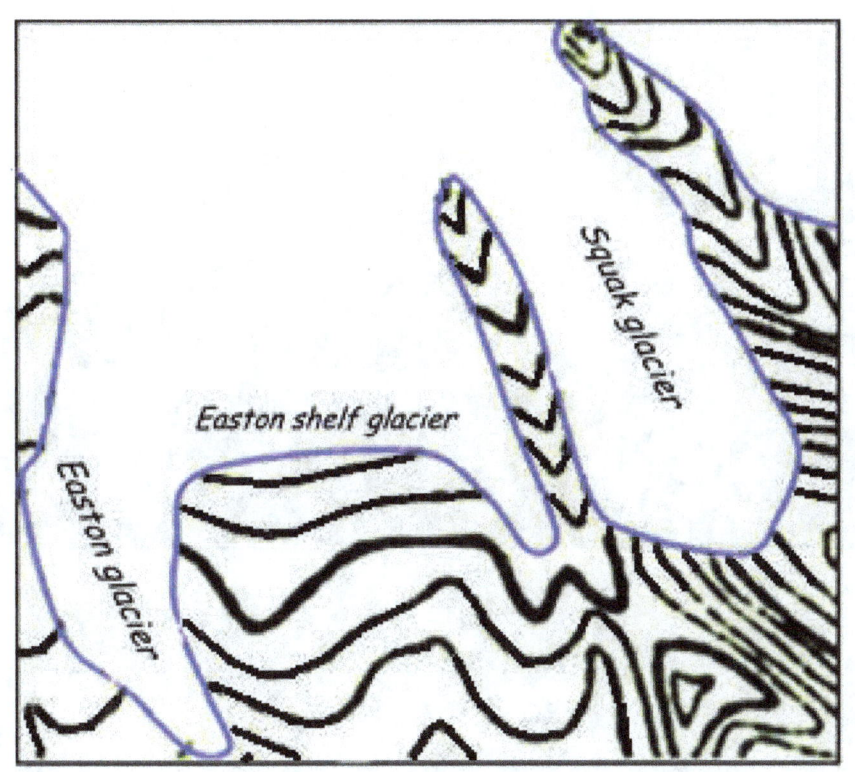

Figure 238. 1909 map of the Easton glacier.

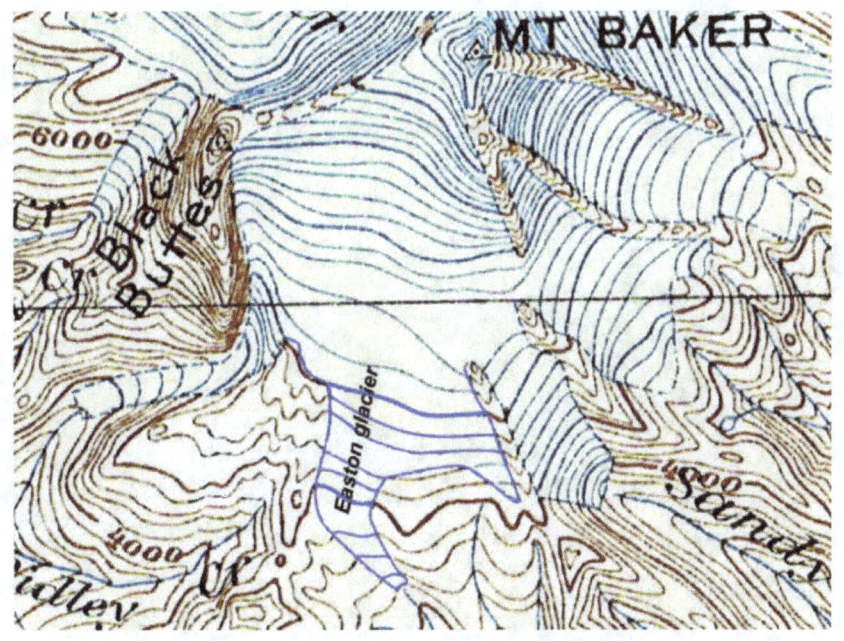

Figure 239. 1915 map of the Easton glacier.

Figure 240. Early photos of the Easton glacier extending far downvalley. [top] 1912, [bottom] early 1900s (Photos by Ed Walsh).

Figure 241. Easton glacier extending far downvalley in the early 1900s. (Photo by Ed Walsh)

Figure 242. Easton glacier, 1930-35.

Figure 243. Easton glacier, 1940.

Figure 244. Easton glacier, 1947.
The terminus retreated 453 feet from 1940 to 1947.

By 1956, the Easton glacier had begun to readvance in response to cooling of the Pacific Ocean offshore. Other glaciers on Mt. Baker had begun to advance earlier, about 1947, but the Easton glacier apparently had a lag response time of about a decade before responding to the change of climate. From 1956 to 1964, the glacier advanced 660 feet (Figs. 248-249) and from 1964 to 1970 it advanced 400 feet. From 1970 to 1979, it advanced 830 feet (Figs. 249-250) and from 1979 to 1985, advanced 144 feet. From 1985 to 1987, the advanced slowed to 7 feet and turned negative from 1987 to 1990, receding 50 feet and continued to retreat for several decades.

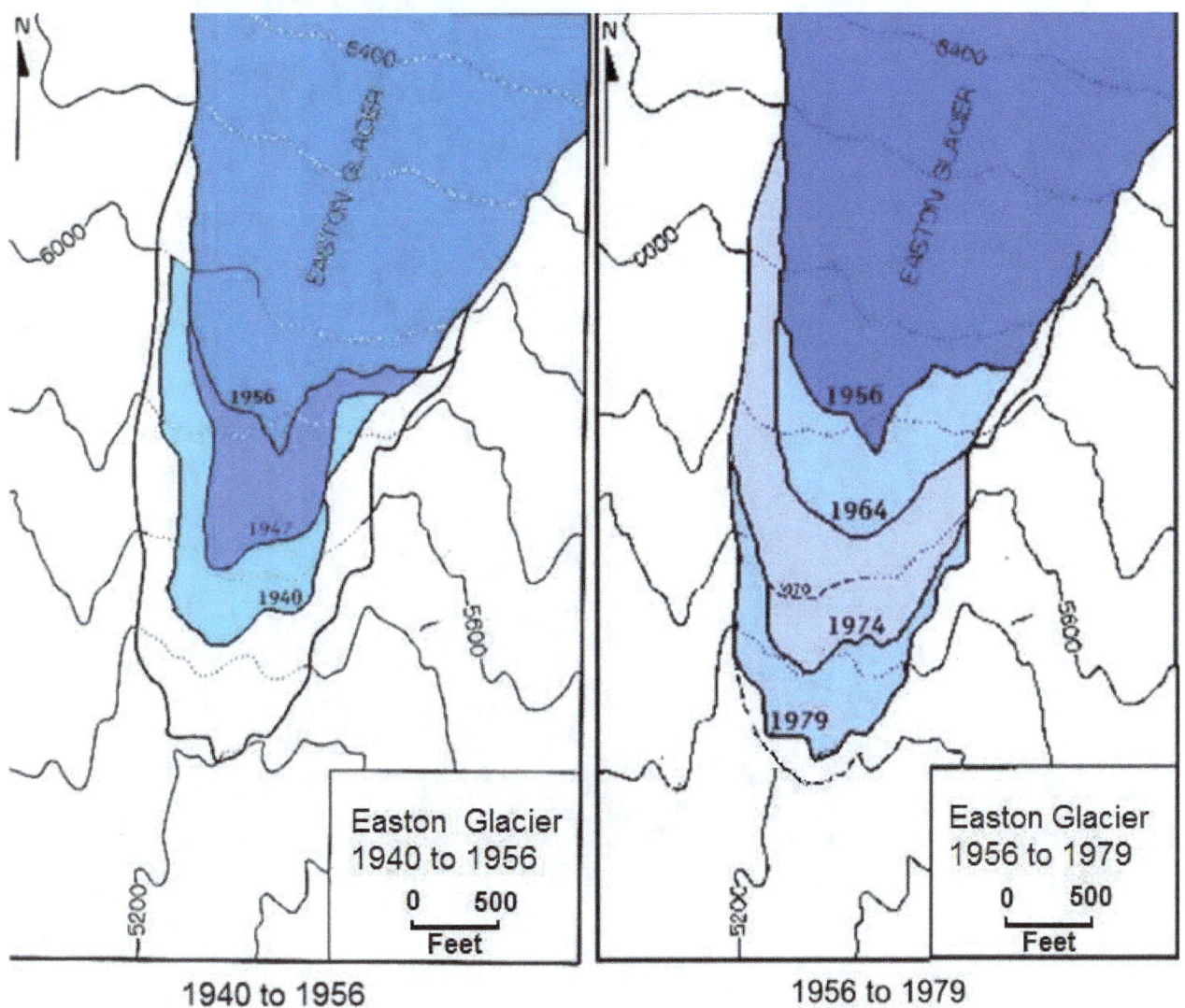

Figure 245. [left] Retreat of the Easton glacier 1940 to 1956.).
[right] Advance of the glacier from 1956 to 1979. (Modified from Harper, 1992

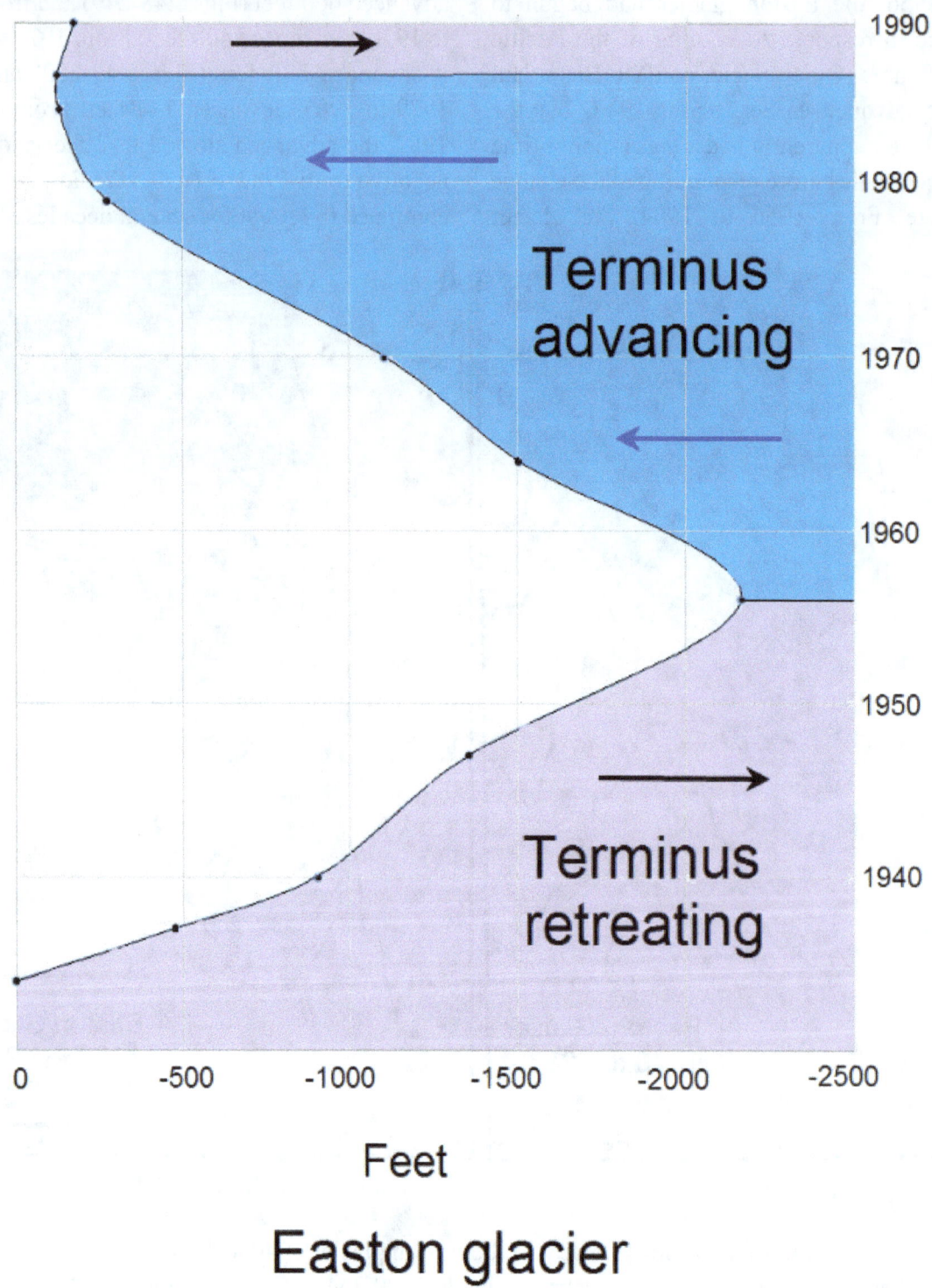

Figure 245B. Amount of retreat and advance of the Easton glacier 1940 to 1990.

Figure 246. Easton glacier 1952. (USGS map)

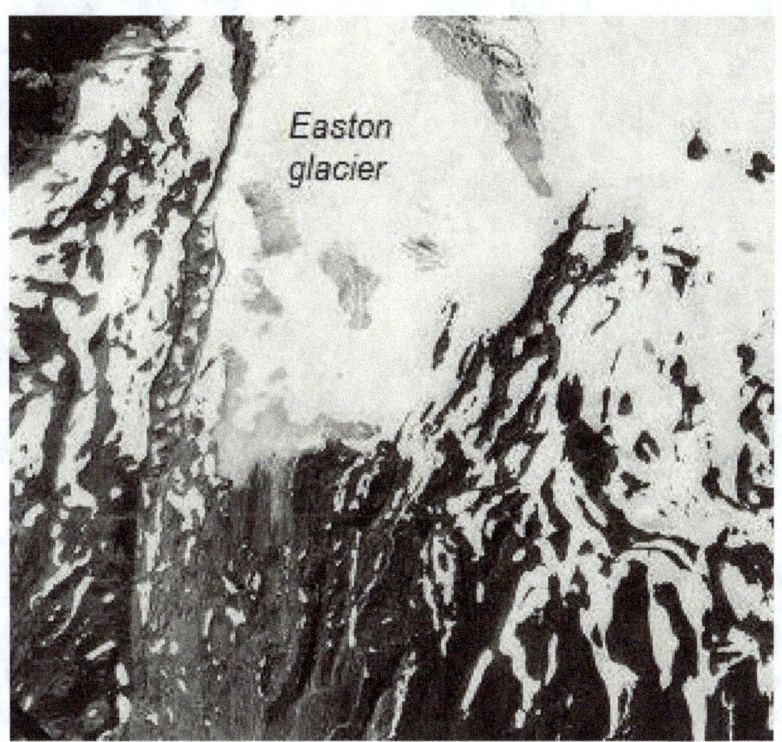

Figure 247. Easton glacier 1950.

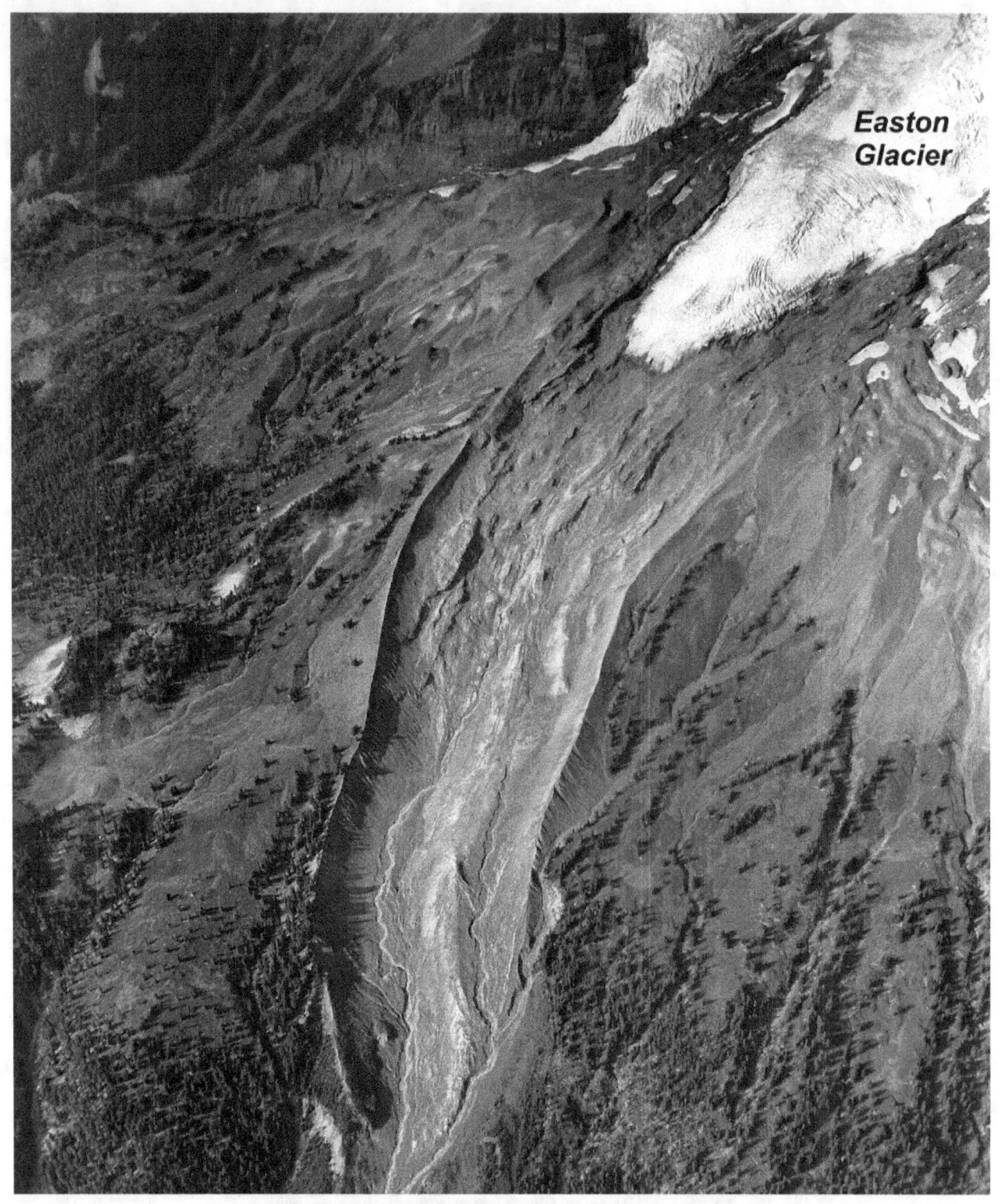

Figure 248. Easton glacier, 1958. (Photo by Austin Post)

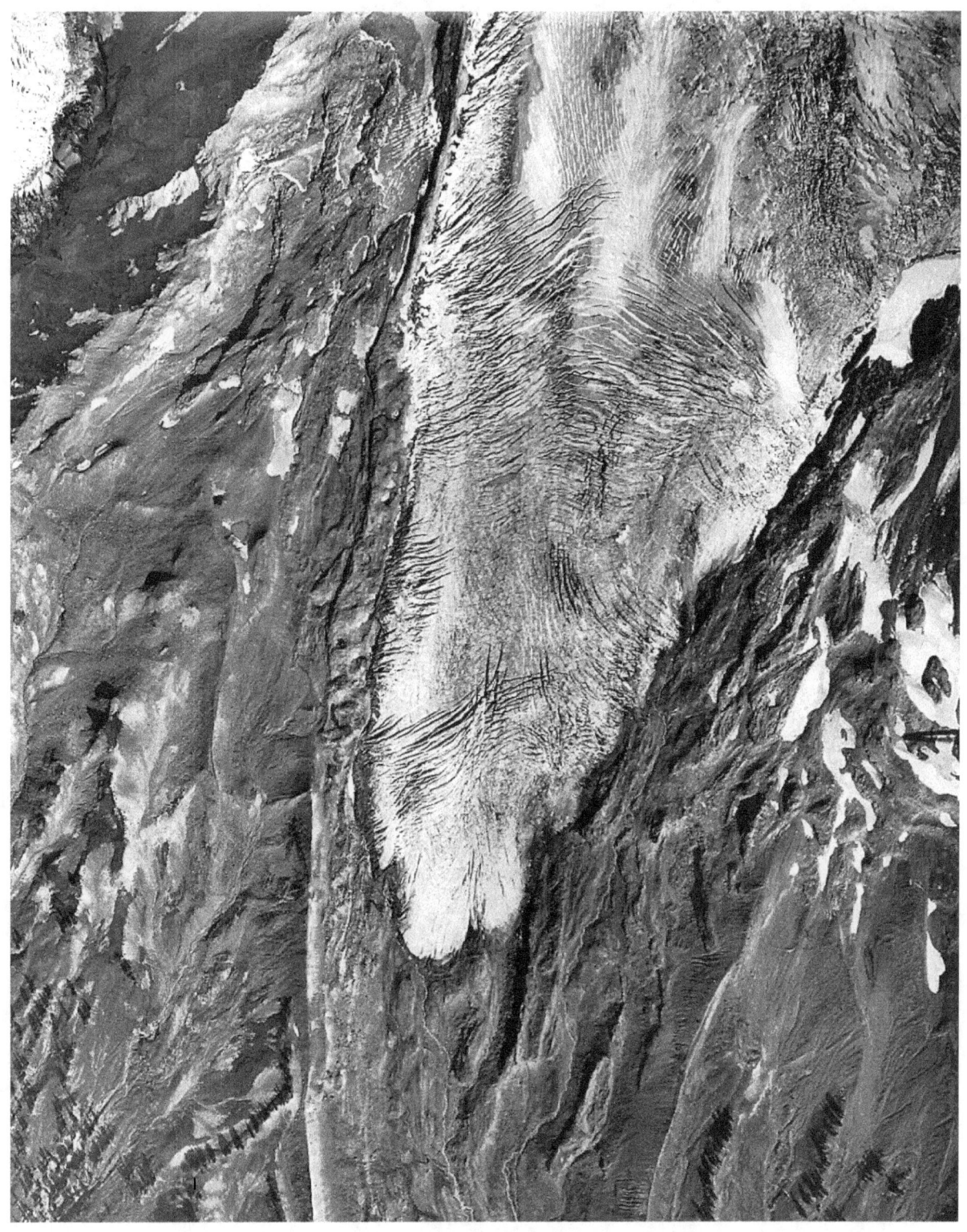

Figure 249. Easton glacier, 1970. (Photo by Austin Post)

Figure 250. Easton glacier, 1979 (Photo by Austin Post)

The Easton lagged behind other Mt. Baker glaciers in responding to the 1979 to 2000 climatic warming. The glacier continued to recede from 1990 to 2014. The terminal position in 2009 is shown in Figure 253.

Figure 251. Easton glacier, 1987. (Photo by Austin Post)

Figure 252. Easton glacier (2002) and lateral moraines. (Photo by Austin Post)

Figure 253. Easton glacier, 2009.

Figure 254 shows the position of the Easton glacier terminus in 1950 and Figure 255 shows the terminus in 2009. The yellow Xs are points of common reference (they are the same place). Yellow diamonds indicate the glacier terminus in 1950 and 2009. These photos show that the Easton glacier is more extensive now than it was in 1950. Figure 256 shows a comparison of the positions of the glacial termini taken from 1952 and 2014 USGS topographic maps. The Easton glacier was more extensive in 2014 than it was in 1952.

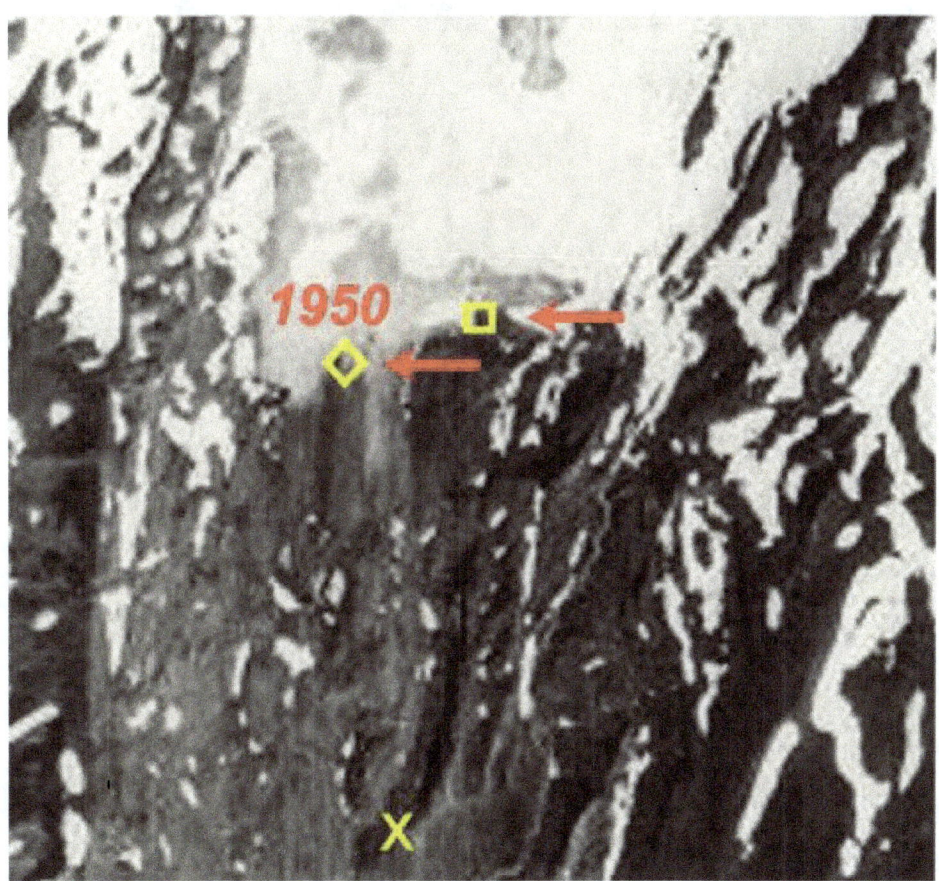

Figure 254. Position of the Easton terminus in 1950.

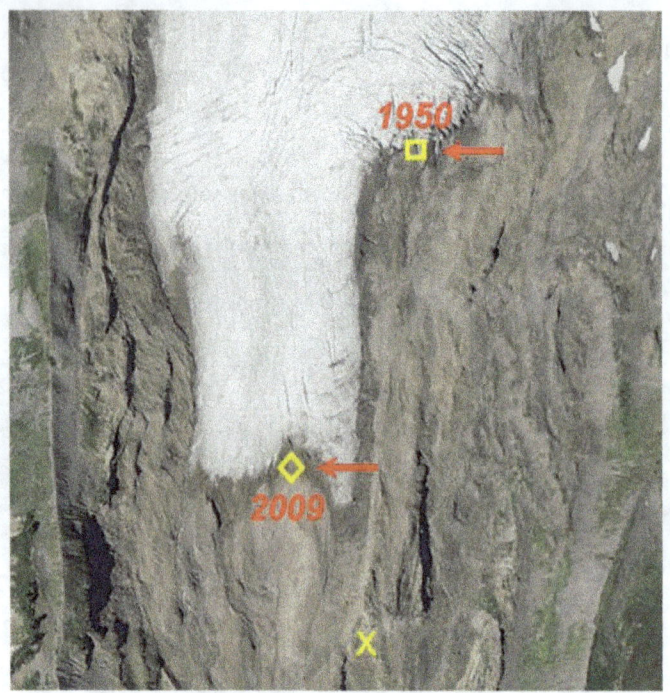

Figure 255. Position of the Easton terminus in 2009.

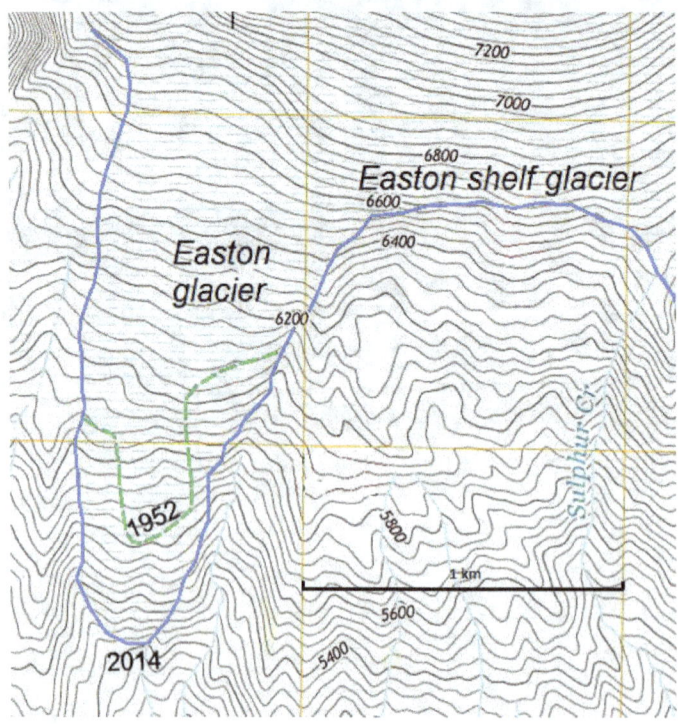

Figure 256. Comparison of the position of the 1952 glacier terminus (green) with the 2014 terminus (blue). The Easton glacier was more extensive in 2014 than it was in 1952.

Easton shelf moraines

Two sets of multiple, digitate moraines extend across the sloping shelf between the Easton and Squak glaciers (Figs. 257-260). A broad band of bouldery, unvegetated moraines rises above older, vegetated, morainal ridges downslope. The ages of these moraines are shown by their relationship to the Scheibers Meadow scoria and Mazama and Rocky Creek volcanic ashes. The age of the older of the two sets of moraines has been much discussed by geologists but remains problematic because the relationship between dated ash layers and moraines is not entirely clear.

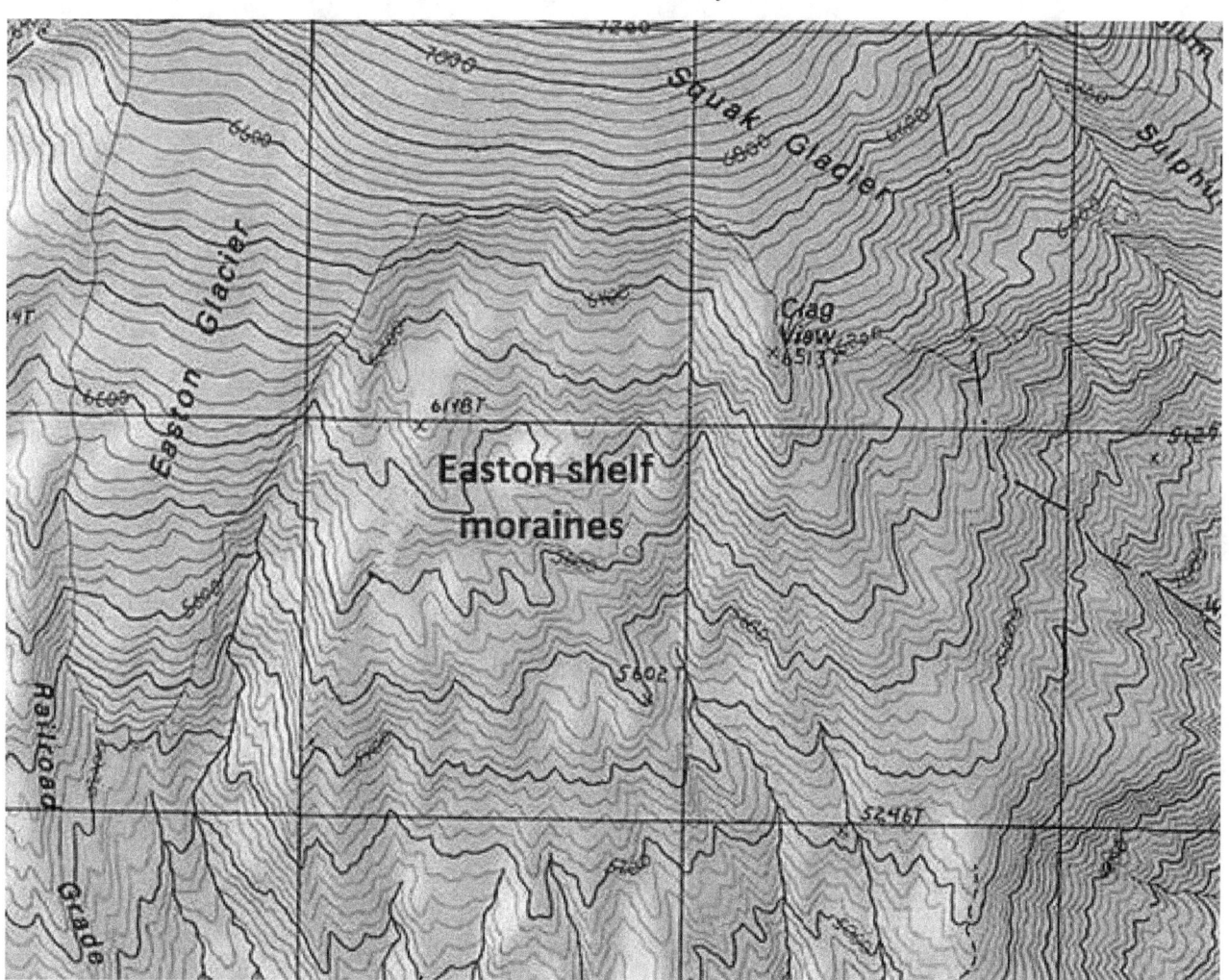

Figure 257. Topographic map of the Easton shelf moraines.

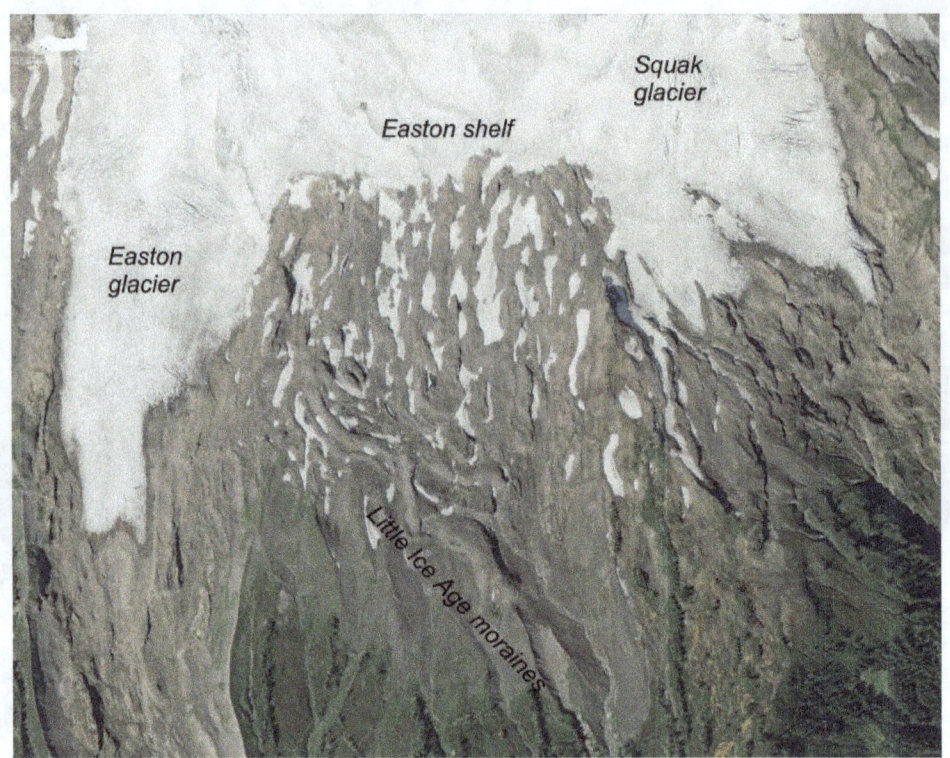

Figure 258. Easton shelf glacier and Little Ice Age (~1500 AD) moraines.

Figure 259. Moraines on the Easton shelf between the Easton and Squak glaciers. (Photo by Austin Post)

Figure 260. Little Ice Age moraines on the Easton shelf.

Younger Dryas/early Holocene and Little Ice Age moraines

A poorly sorted mixture of silt, sand, pebbles, and cobbles on west side of the Easton shelf contains wood dated at 11,020 ± 80 ^{14}C yrs (Fig 261), placing it within the Younger Dryas cool period. Peat overlying glacial deposits on the eastern side of the shelf has been dated at 11,460 ± 35 ^{14}C yrs, suggesting that the underlying glacial deposits are pre–Younger Dryas in age.

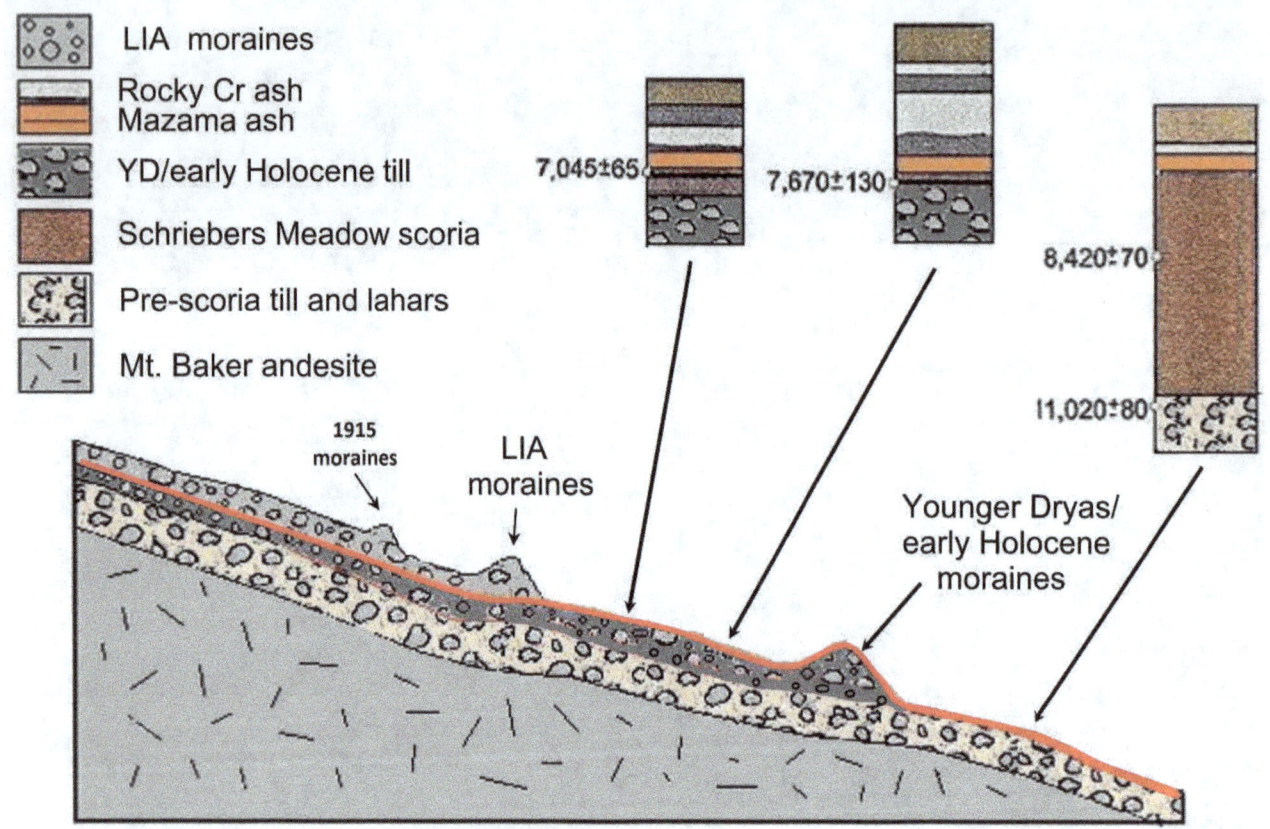

Figure 261. Geologic cross-section of the Easton shelf moraines.

The older of the two sets of moraines on the Easton shelf consists of multiple ridges with trees growing along the crests (Figs. 261-265) and mantled with Schreibers Meadow scoria and Mazama ash. The moraines must be older than the scoria (8,800 ^{14}C years old), so are between 9,000 and 12,500 ^{14}C years old, i.e., Younger Dryas and/or early Holocene. The age of the Pocket Lake moraine just west of Railroad Grade contains wood dated at 8,400 years (see below), but Pocket lake sediments behind the moraine contain organic material dated at 11,455 ± 110 ^{14}C years, with only a few inches of sediment deposited until 7,640 ± 50. The cirque was apparently filled with ice between 11,500 and 7,600 years ago, preventing deposition of sediment in the lake. Thus, the moraine was apparently built during two phases, an early Younger Dryas phase and a later early Holocene phase.

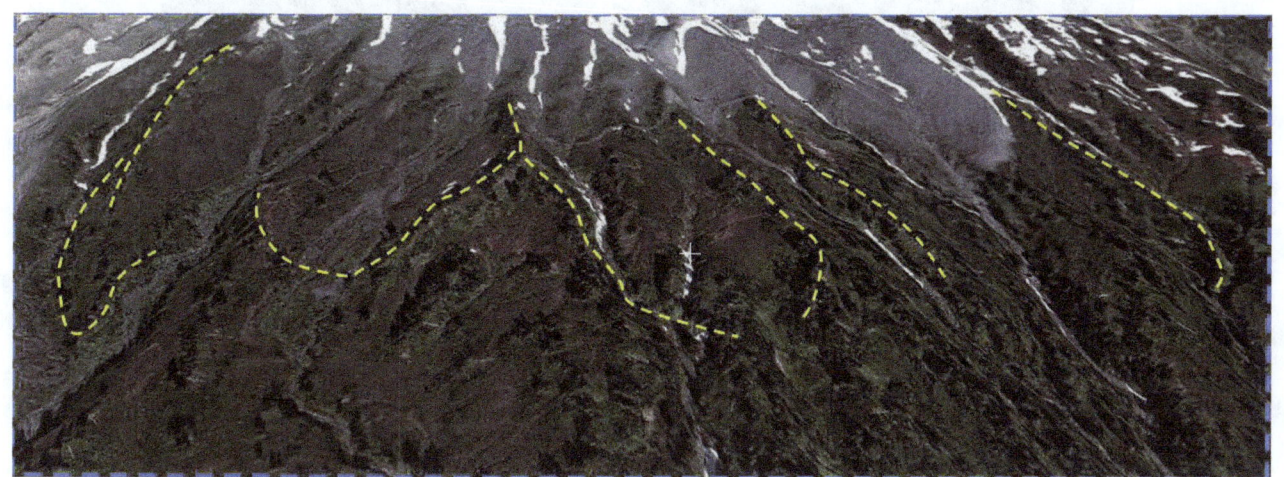

Figure 262. Older of the two sets of moraines on the Easton shelf (dashed lines).

Figure 263. Outer Easton shelf moraines. The long gray ridge near the top is Railroad Grade, a lateral moraine of the Easton glacier. Park Butte is on the upper left and the Twin Sisters Range makes the central skyline.

Figure 264. Tree covered crests of the older set of Easton shelf lateral moraines.

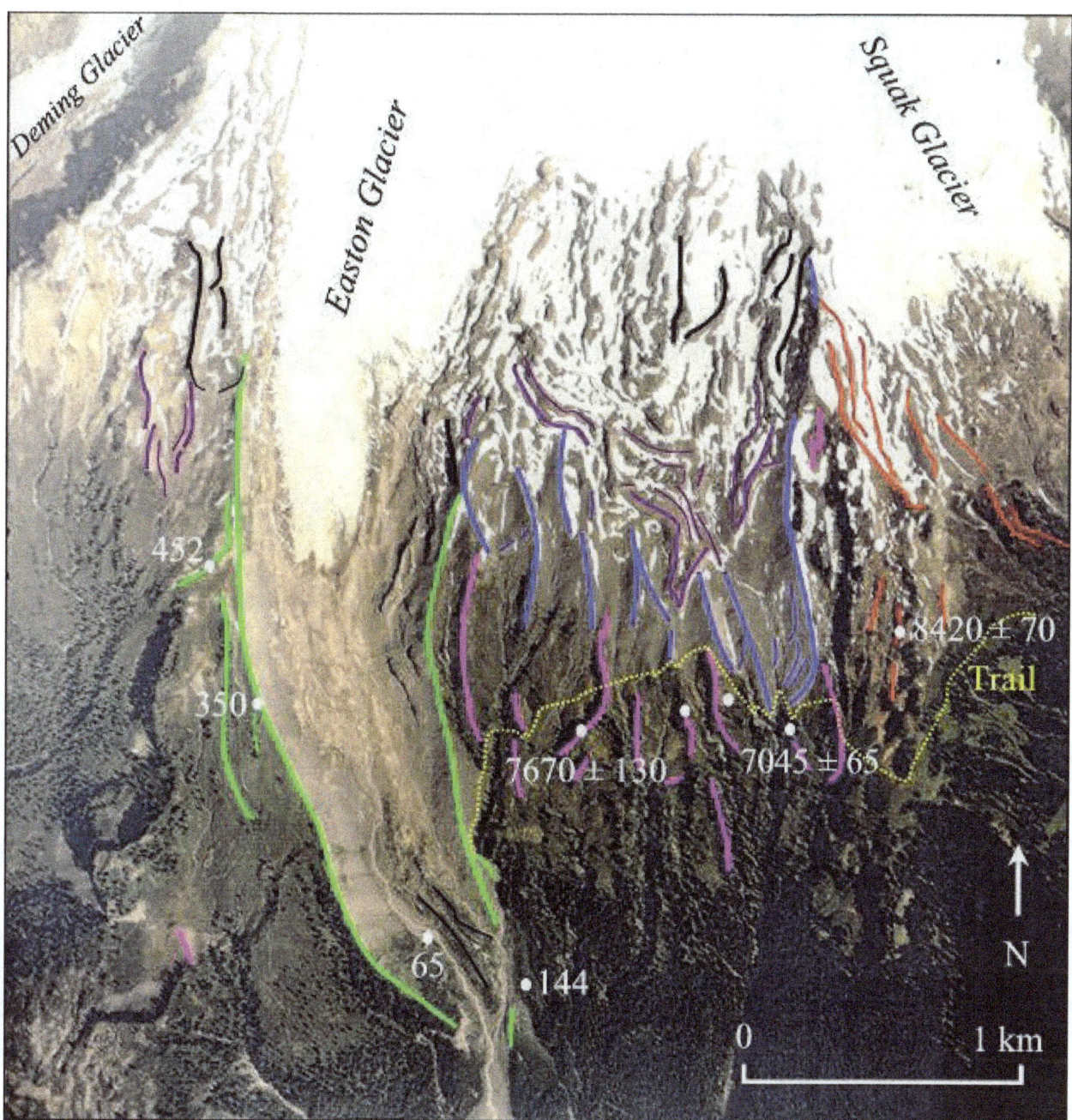

Figure 265. Moraines on the Easton shelf. The older of two sets of moraines is shown by Vermillion lines; numbers are ^{14}C dates of organic material on the moraines. The outermost moraines of the younger set of moraines are shown in blue. Numbers along the west side of the valley of the Easton glacier are ages of the largest trees growing on the Railroad Grade moraine.

Figure 266. Mazama ash beneath glacial deposits on the Easton shelf.

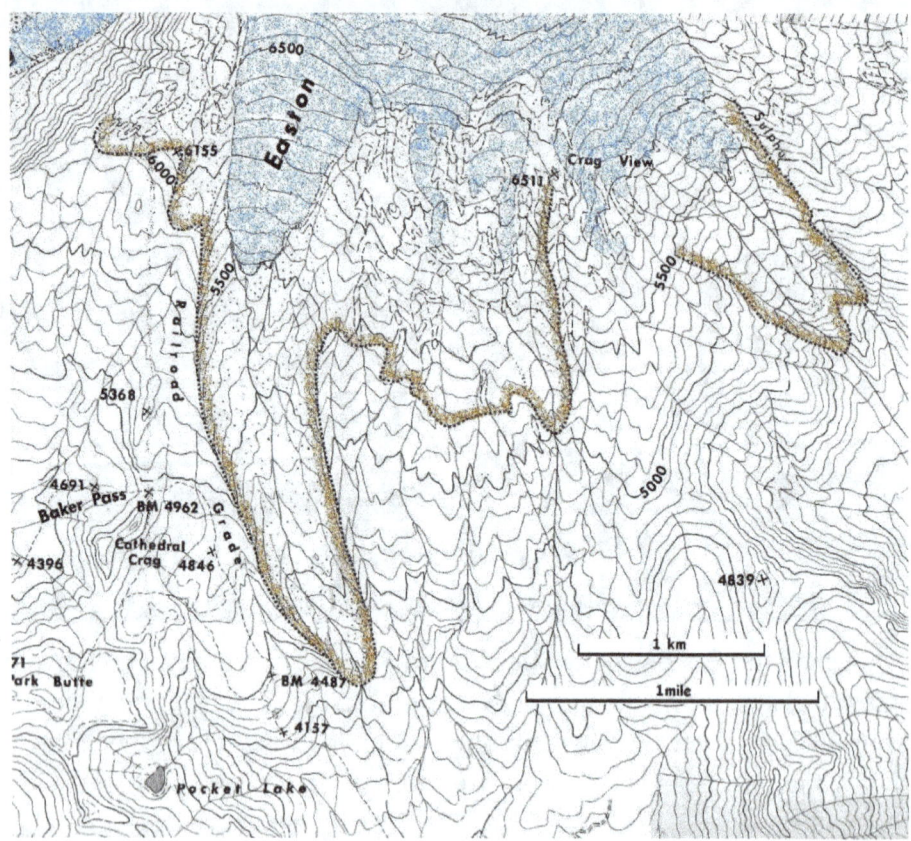

Figure 267. Little Ice Age ice margins (brown pattern) across the Easton shelf.
(Modified from USGS topographic map)

Little Ice Age moraines

The younger of the two sets of moraines on the Easton shelf consists of a long, bouldery, treeless ridge with strongly digitate margins (Figs. 258-260; 268, 269). The age of the moraines is younger than Schreibers Meadow scoria (8,800 ^{14}C years) and Mazama ash (6,800 ^{14}C years). Although no finite dates are available from the moraines, they were most likely deposited during the Little Ice Age (1300 to 1915 AD).

The youngest of the Little Ice Age moraines consist of three inner moraines (Fig. 269) draped across bedrock knobs. At least three small moraines are draped over bedrock behind the outer Little Ice Age moraines (Fig. 269). These moraines appear to mark the 1915 ice margin. The 1915 USGS topographic map shows the ice margin on the Easton shelf at the same elevation as the inner Little Ice Age moraines.

Figure 268. Little Ice Age moraines on the Easton shelf.

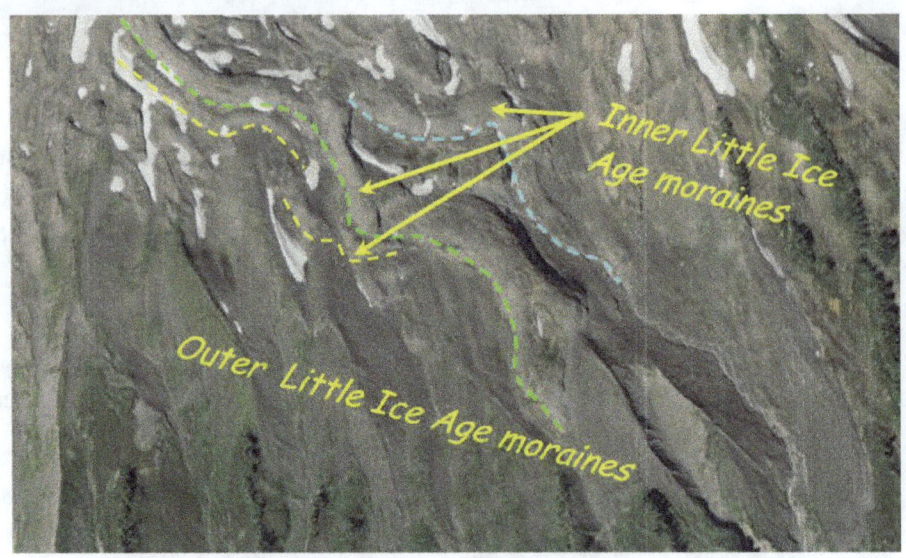

Figure 269. Little Ice Age moraines on the Easton shelf.

The maximum size of the Easton shelf glacier was attained about 1915 at the end of the 1880-1915 cool period (Figs. 270, 271). During the 1915-1945 warm period, the glacier receded strongly (Fig. 272). Figure 274 shows its extent in 1952. During the following cool period (~1945 to 1979), the glacier readvanced but not as far downvalley as it had been in 1915.

The climate turned warm again from about 1980 to 2000, but even though this retreat of the terminus did not begin as far downvalley as the 1915-1952 recession, by 2014, the ice had not yet receded as far as it did in 1952. Thus, the glacier is still more extensive now than it was in 1952 (Fig. 276).

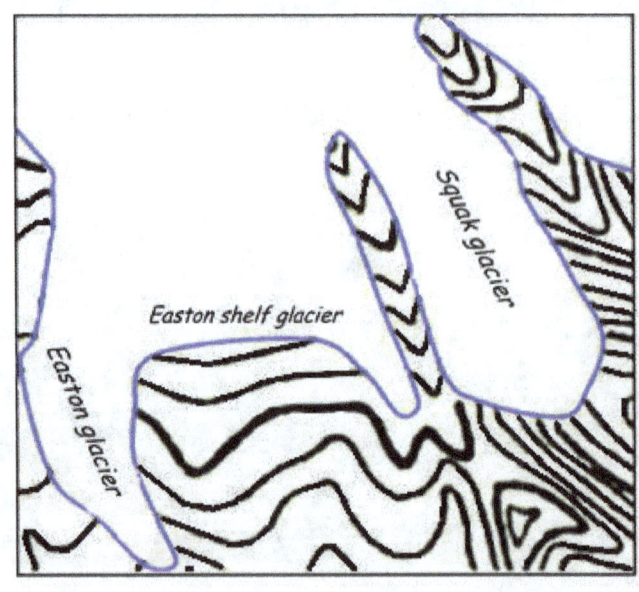

Figure 270. Easton shelf terminus, 1909. (USGS map)

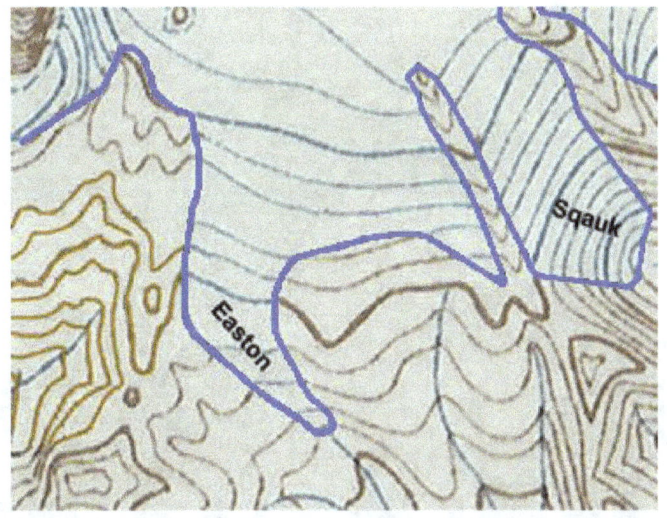

Fig. 271. Easton shelf terminus, 1915 margin. (USGS map)

Figure 272. Easton shelf glacier, 1940.

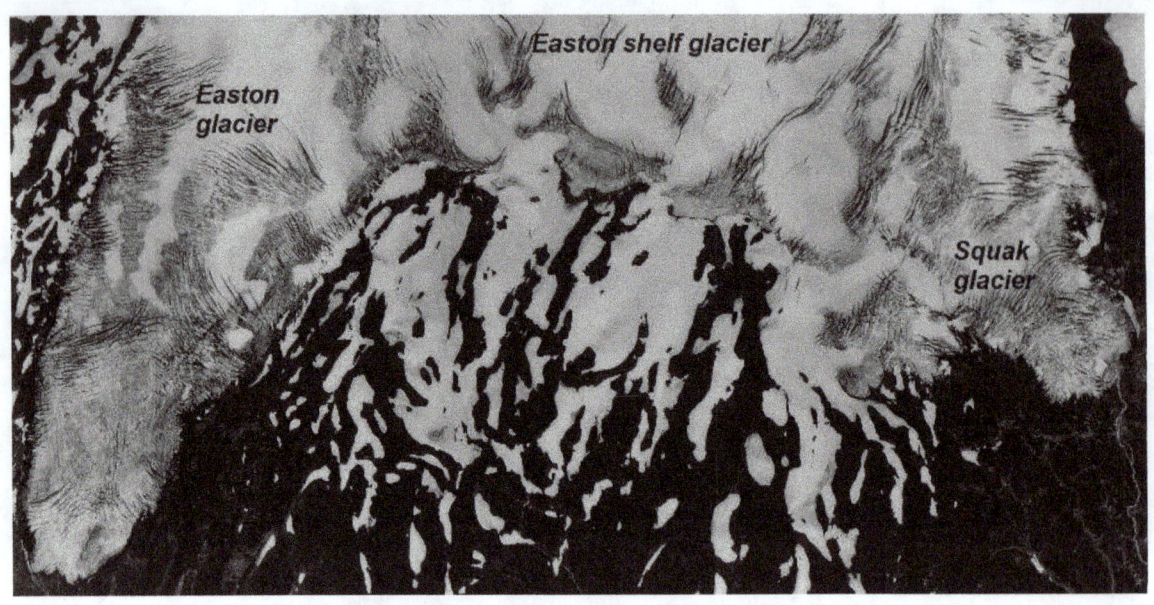

Figure 273. Easton shelf glacier, 1985.

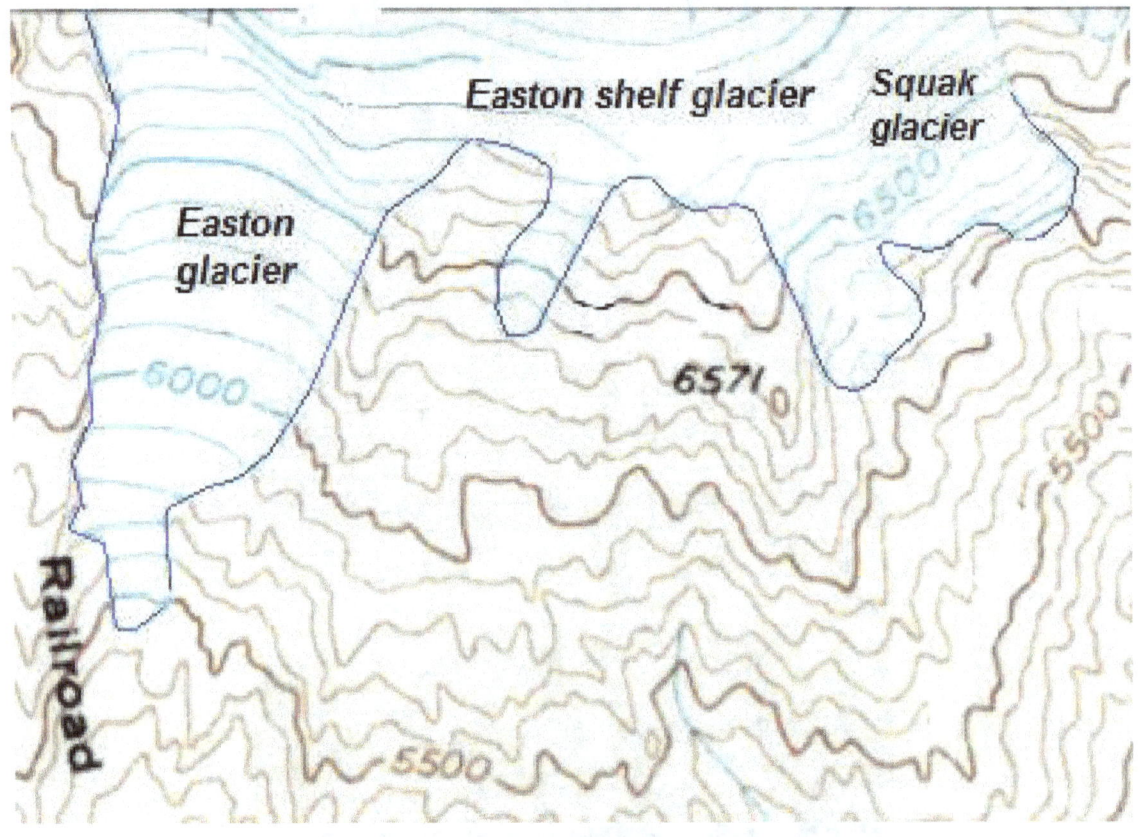

Figure 274. Easton shelf glacier, 1952.

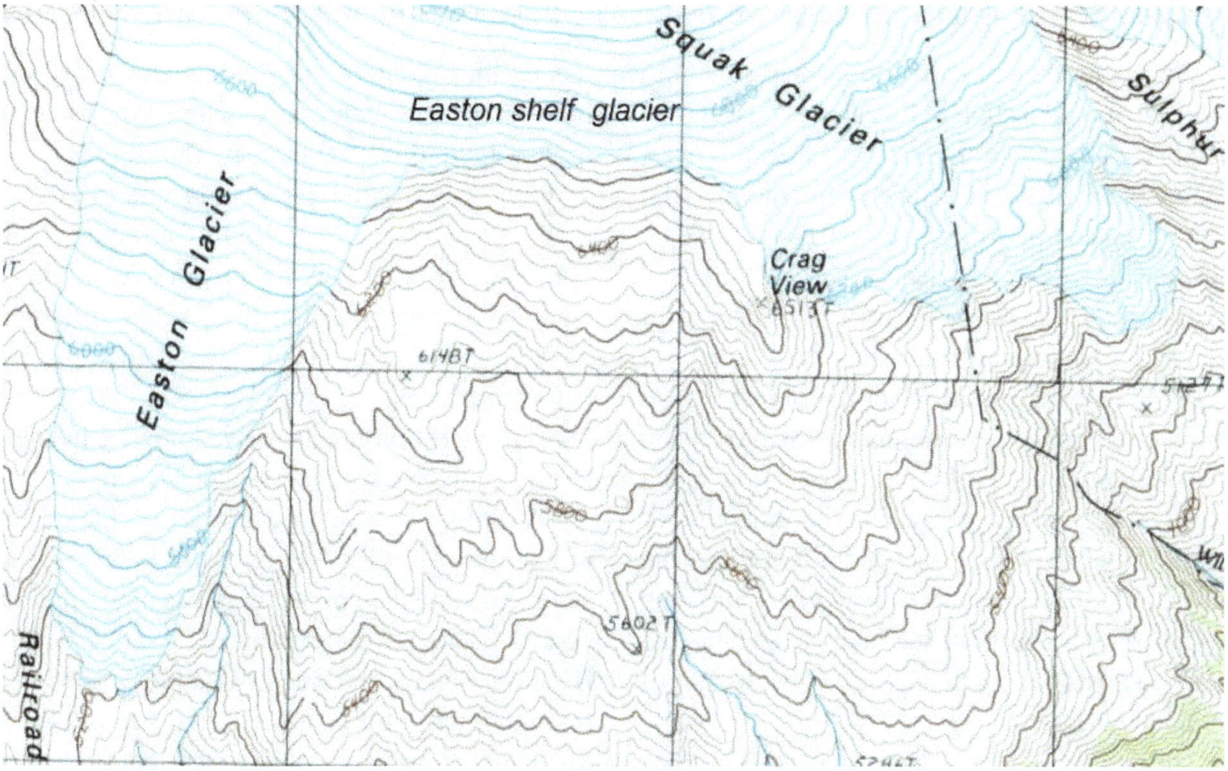

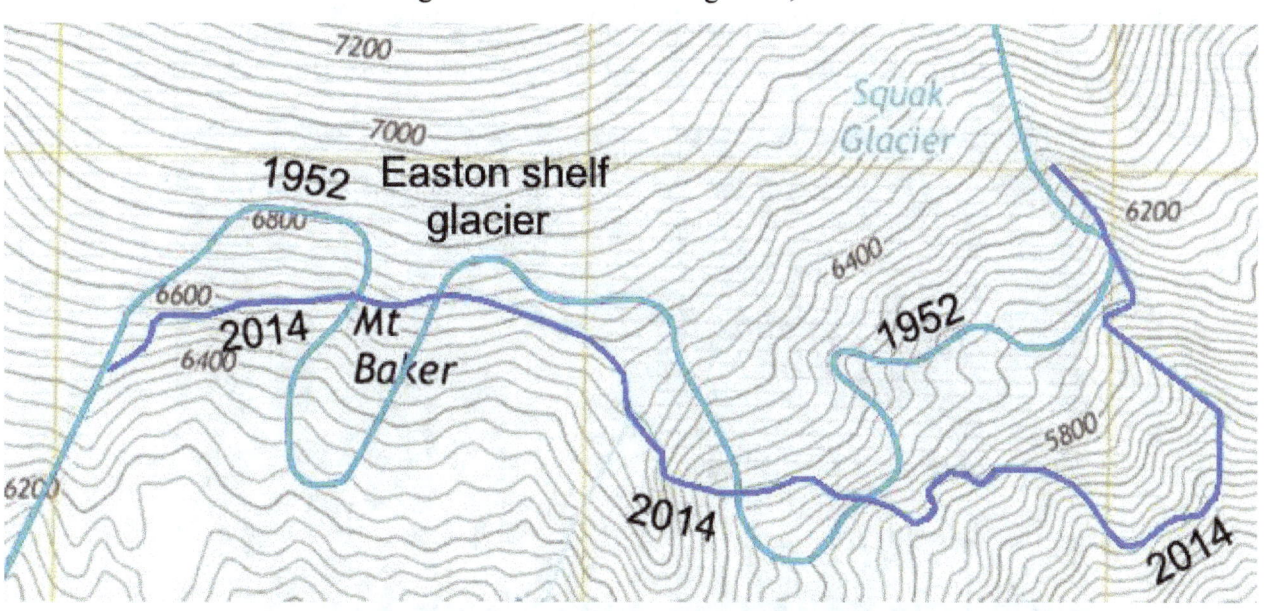

Figure 275. Easton shelf glacier, 2014.

Figure 276. Comparison of Easton glacier in 1952 (green line) and 2014 (blue). The glacier was more extensive in 2014 than it was in 1952.

Younger Dryas/Early Holocene moraines at Pocket Lake and Park Butte cirques

The occurrence of early Holocene (10,000 to 8000 ^{14}C yrs) glaciation in this area is best shown at Pocket Lake and Park Butte cirques just west of the Easton glacier (Fig. 277). Organic material in the Pocket Lake moraine has been dated at 8,455 ± 75 ^{14}C yrs, clearly early Holocene. Organic matter at the base of a 4–foot core of Pocket Lake bottom sediments was dated at 11,455 ± 110. This means that the oldest part of the moraine holding in the lake must have been deposited prior to 11,455 ^{14}C yrs. How much before that is unknown because the thickness of sediment between the base of the core and the bedrock floor of the cirque is not known. The 11,455 ^{14}C yrs age from the lake silt means that the cirque was free of glacial ice at beginning of the Younger Dryas.

The lake silt at the bottom of the core is overlain by three inches of organic silt. A radiocarbon date of 7,640 ± 50 from the organic silt three inches above the underlying lake silt means that 3,800 years passed with only three inches of sedimentation. This time interval includes all of the Younger Dryas (11,500 to 10,000 ^{14}C yrs), the early Holocene (10,000 to 7,600 ^{14}C yrs), and the time of eruption of the Schreibers Meadow scoria. Organic matter in the Pocket Lake moraine has been dated at 8,455 ± 75 ^{14}C yrs so at least part of the moraine was deposited during the early Holocene. Whether or not the entire moraine is early Holocene in age or has a Younger Dryas core is not known.

A similar moraine occurs at the Park Butte cirque just north of Pocket Lake where lake sediments and ice contact deposits along the side of the basin have been dated at 8820 ± 110. Because there is no plausible cause for impoundment of a lake there (the site is open to the adjacent meadows), these sediments strongly suggest an ice–marginal environment from a glacier in the Park Butte cirque, confirming early Holocene glaciation of the area.

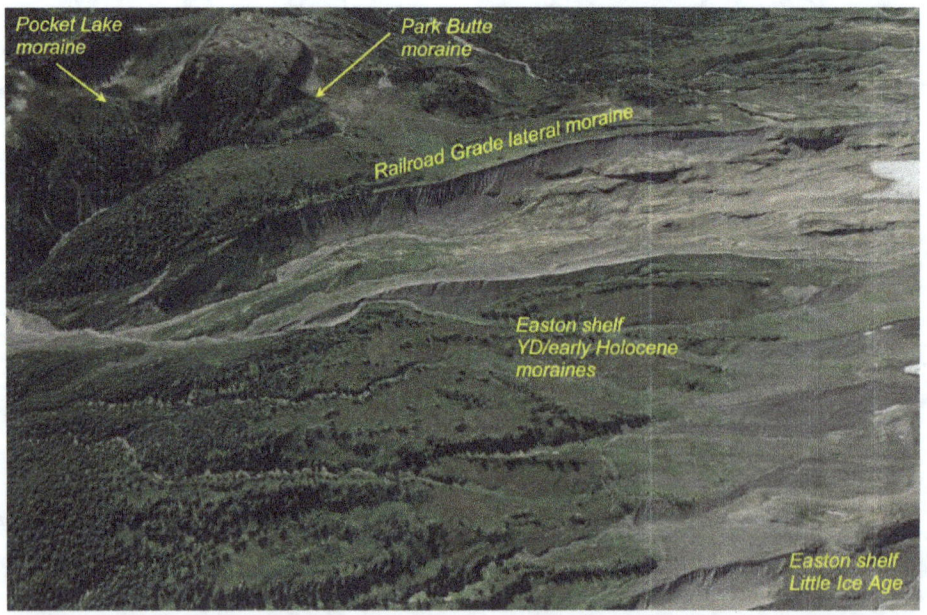

Figure 277. Pocket Lake, Park Butte, and Easton shelf Younger Dryas/early Holocene moraines.

Squak glacier

The Squak glacier originates on the southwest flank of Sherman Peak on Mt. Baker and flows into the upper drainage of Sandy Creek (Fig. 278). It is separated from the Talum glacier on the east side by a ridge known as Sulphur moraine. It merges to the west with ice from the Easton glacier.

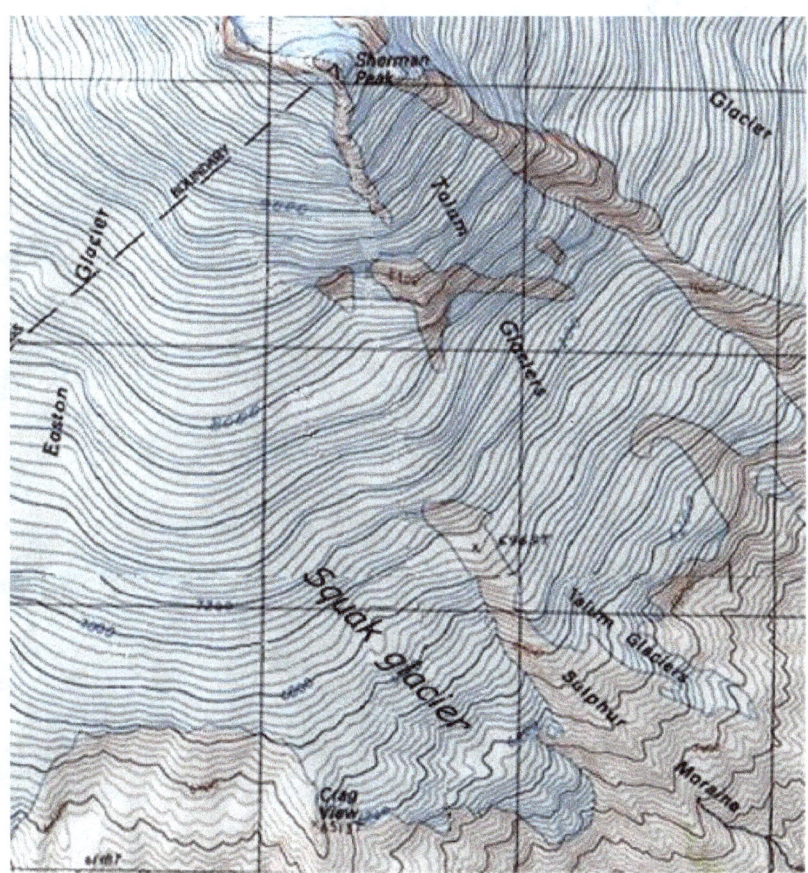

Figure 278A. Topographic map of the Squak glacier.

Figure 278B. Photo of the Squak glacier. (Photo by Austin Post)

Old moraines

Lateral moraines high on the valley sides extend well downvalley from the present Squak glacier terminus (Figs. 279,280). They have not yet been studied or dated so little is known about them. The highest lateral moraines along the valley sides were most likely deposited during the Younger Dryas and Little Ice Age, much like the other glaciers on Mt. Baker.

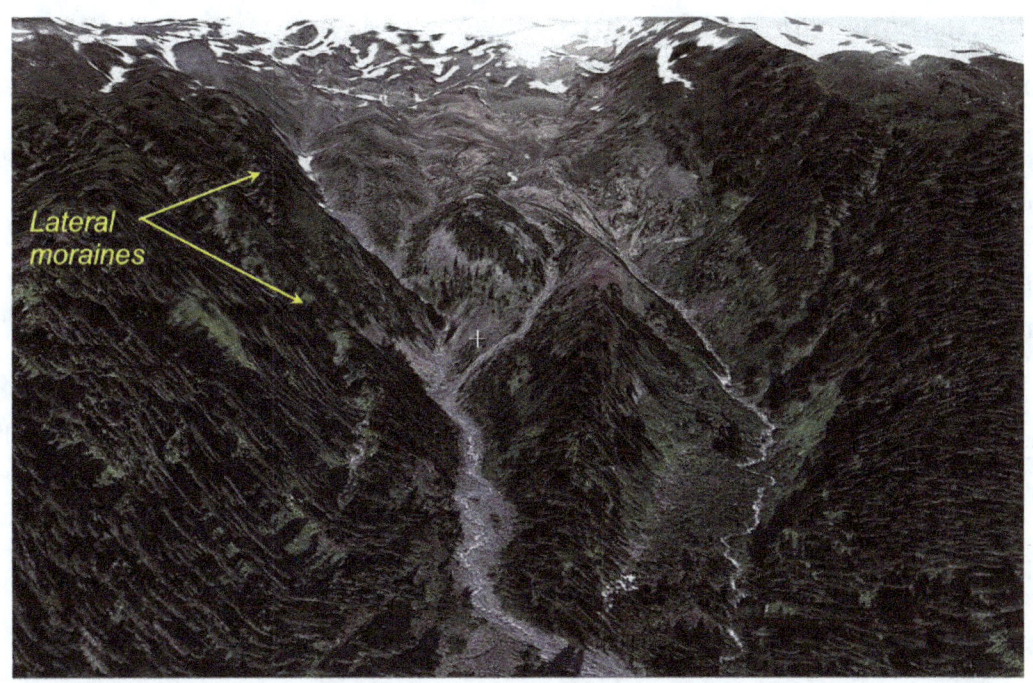

Figure 279. Younger Dryas/Little Ice Age lateral moraines below Squak glacier.

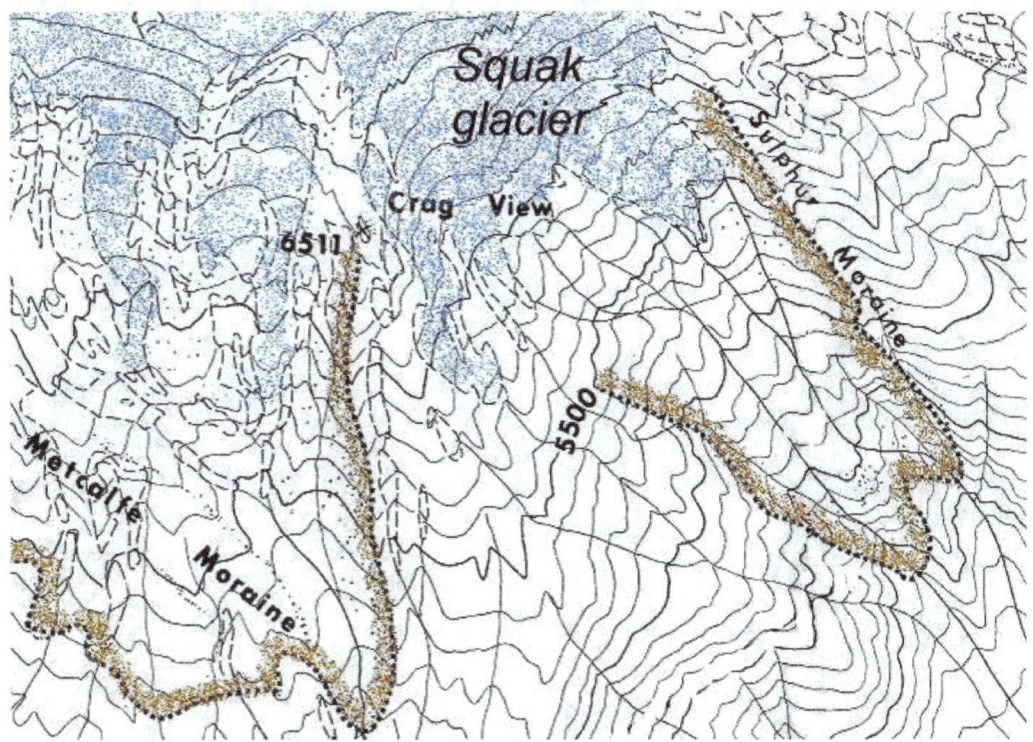

Figure 280. Little Ice Age (~1500 AD) moraines (brown) of the Squak glacier.
(Modified from USGS topographic map)

Glacier fluctuations from 1915 to 2015.

The maximum extent of the Squak glacier in the past century occurred at the end of the 1880–1915 cool period. Figure 281 shows the position of the glacier terminus in 1909 and Figure 282 shows the 1915 terminus. Like other Mt. Baker glaciers, the Squak glacier advanced and retreated as the climate warmed and cooled during the century. From 1915 to about 1945-1950, the warmest climatic conditions of the century caused the Squak glacier to retreat far upvalley to its highest terminal position of the century (Fig. 283).

The climate turned cool between about 1945 and 1950, halting the retreat of the terminus and during the ensuing cool period from about 1950 to 1977, the Squak glacier advanced vigorously (Figs. 284–286), reaching its maximum extent about 1979 (Fig. 286).

The Pacific Ocean flipped abruptly from its cool mode to its warm mode in 1978 and the climate remained warm until about 2000. Although the climate has cooled in the past decade, the Squak glacier continued to retreat until sometime between 2005 and 2009 (Figs. 287–289). Since 2013 the glacier has stopped retreating and remains in about the same position in 2015.

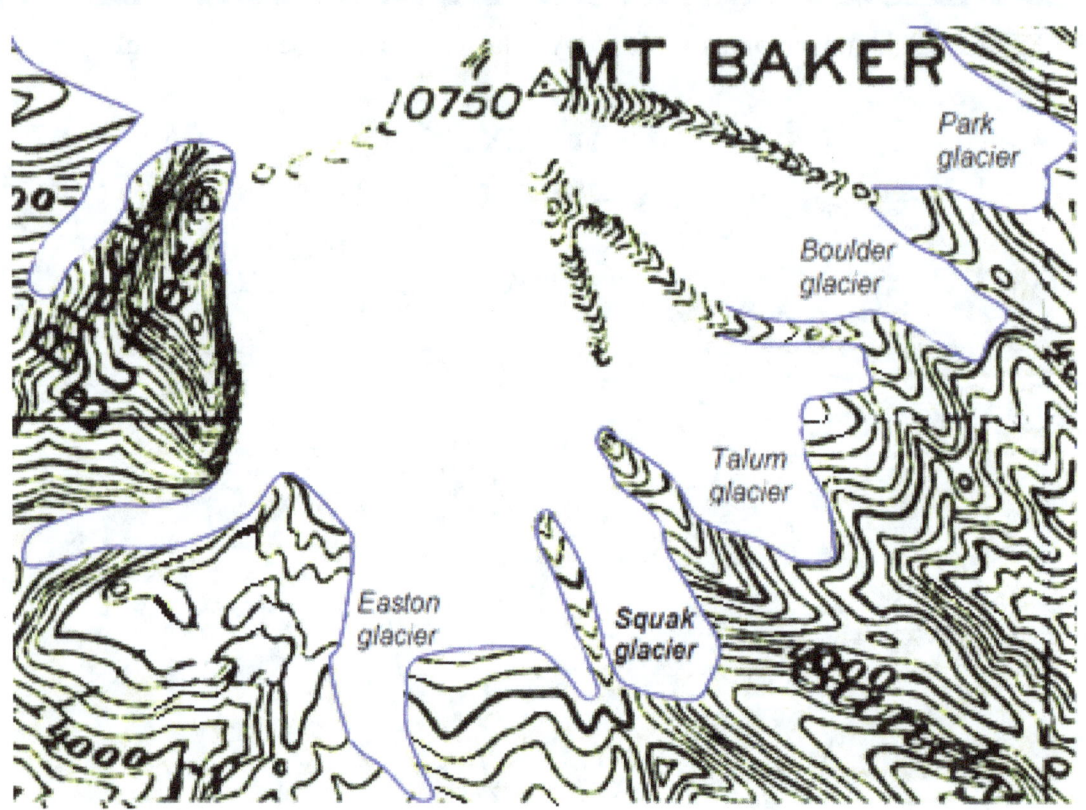

Figure 281. Squak, glacier 1909. (USGS map).

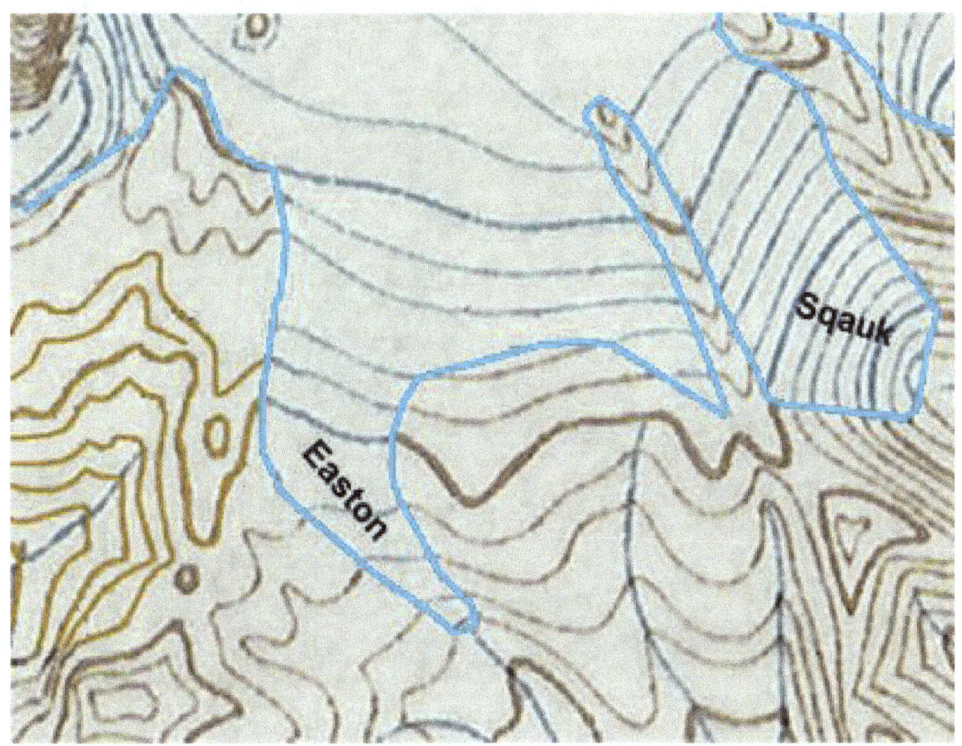
Figure 282. Squak glacier, 1915. (USGS map)

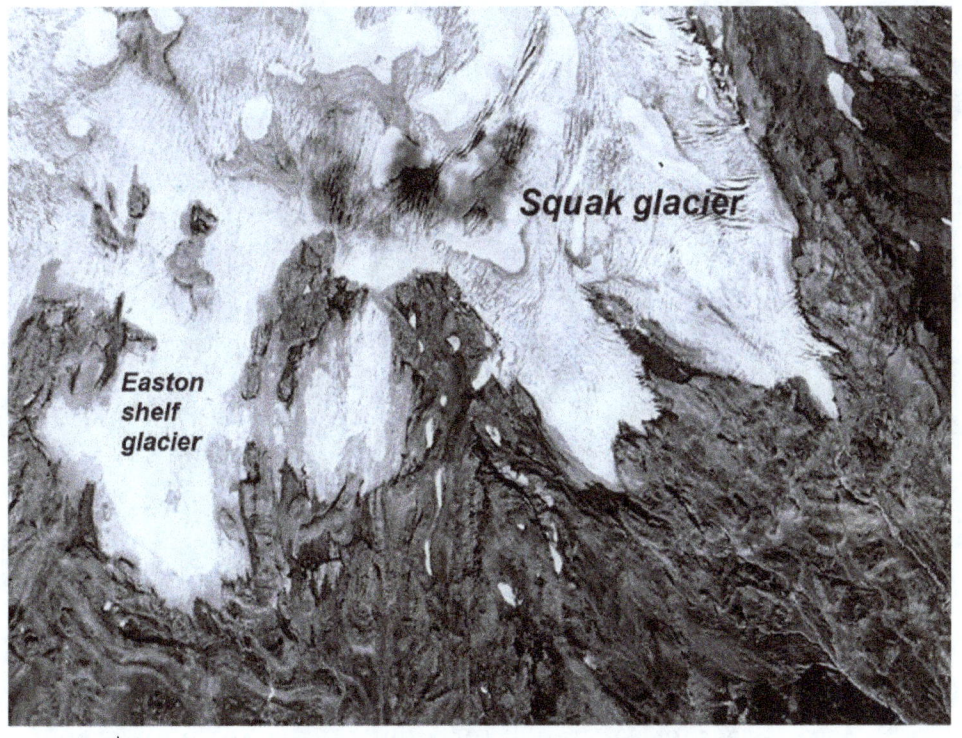
Figure 283. Squak glacier, 1940.

Figure 284. Squak glacier, 1958. (Photo by Austin Post)

Figure 285. Squak glacier, 1965. (Photo by Austin Post)

Figure 286 Squak glacier, 1979. (Photo by Austin Post)

Figure 287. Squak glacier, 1987. (Photo by Austin Post)

Figure 288. Squak glacier.

Figure 289. Squak glacier 2009.

Although the Squak glacier retreated from 1979 to about 2005, it did not retreat as far upvalley as it did from 1915 to 1952. Figure 290 shows the terminus position in 1952 and Figure 291 shows it in 2014. Comparison of the two terminal positions (Fig. 292) shows that the terminus today is about 1500 feet downvalley from the 1952 positon.

Figure 290. Squak glacier, 1952. (USGS map)

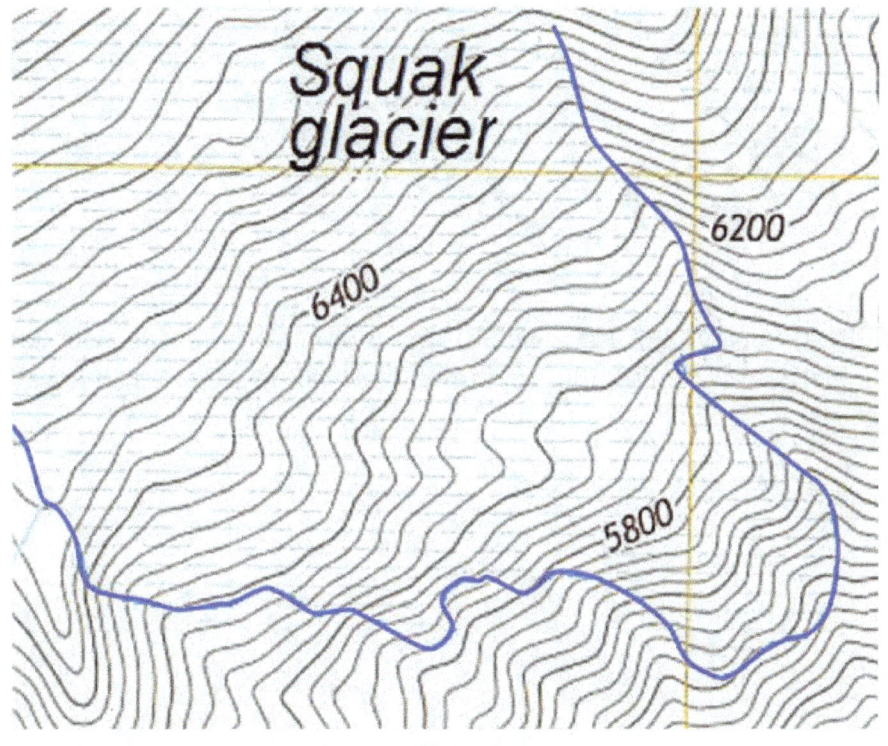

Figure 291. Squak glacier, 2014. (USGS map)

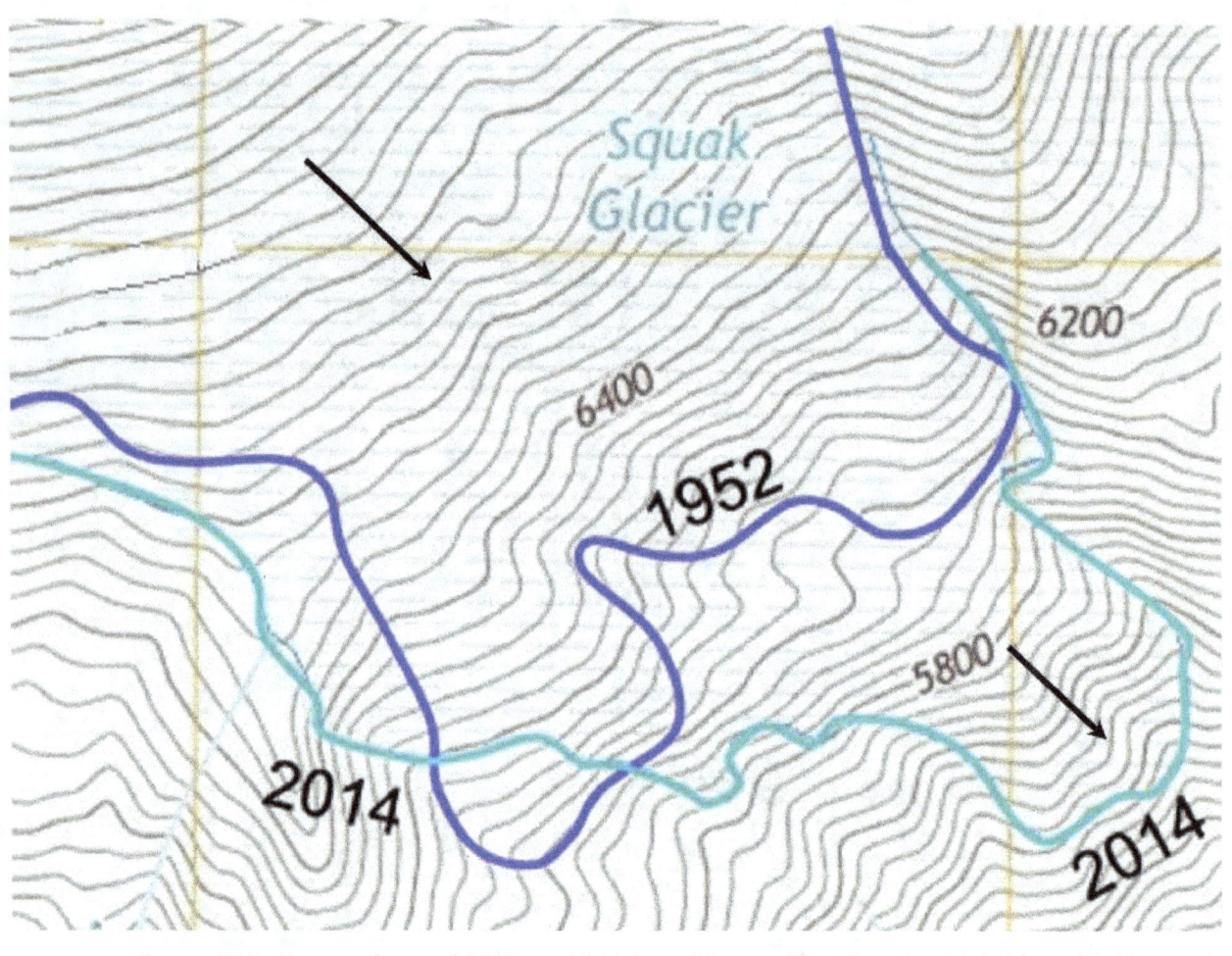

Figure 292. Comparison of 1952 and 2014 positions of the Squak glacier terminus. The Squak glacier was about 1,500 feet farther downvalley in 2014 than it was in 1952.

Talum glacier

The Talum glacier originates from the SW side of Sherman Peak and makes a broad expanse of ice that splits into three lobes at the ice terminus (Figs. 293–295).

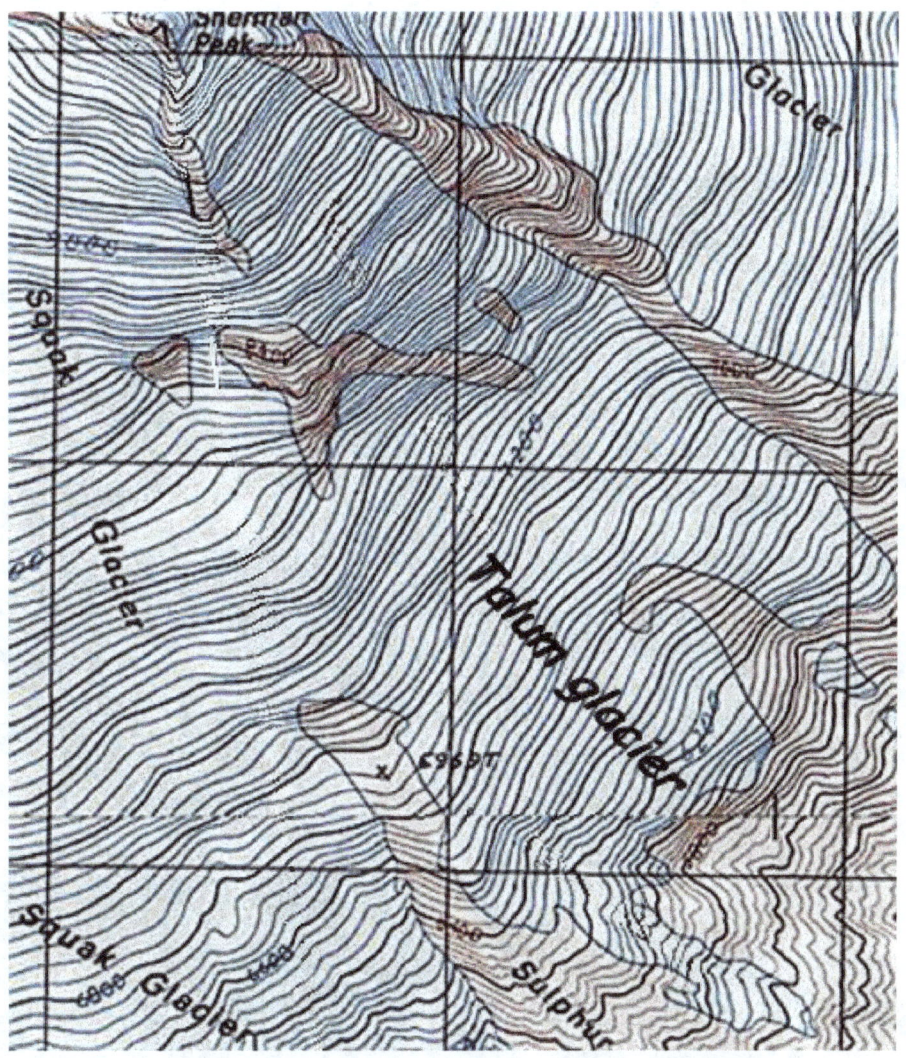

Figure 293. The Talum glacier. (USGS map)

Figure 294. The Talum glacier. (Photo by Austin Post)

Figure 295. Talum glacier.

Little Ice Age and 1915 ice margins

The Talum glacier extended much farther downvalley during the Little Ice Age (~1500AD) and built lateral moraines high along the valley sides (Figs. 295, 298). The strongest advance of the Talum glacier occurred at the end of the 1880–1915 cool period when the terminus extended the farthest downvalley during the past century (Fig. 298) and the strongest recession occurred from 1915 to ~1952.

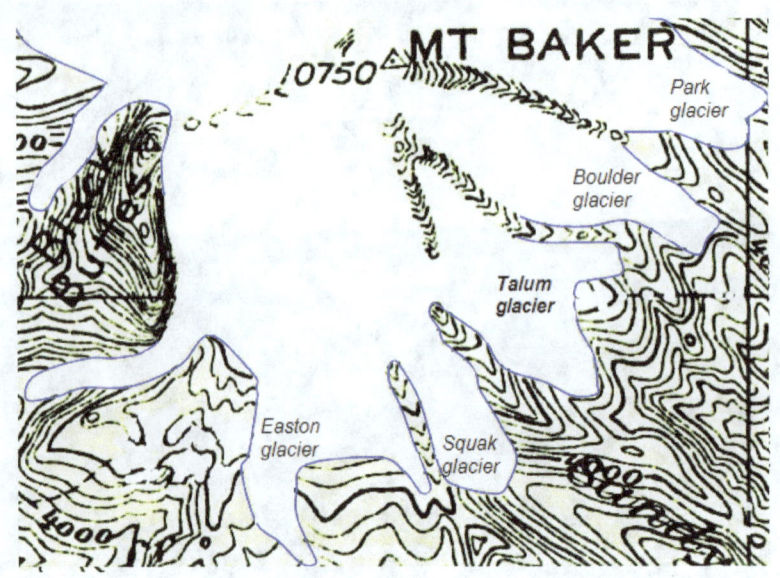

Figure 296. Talum glacier, 1909. (USGS map)

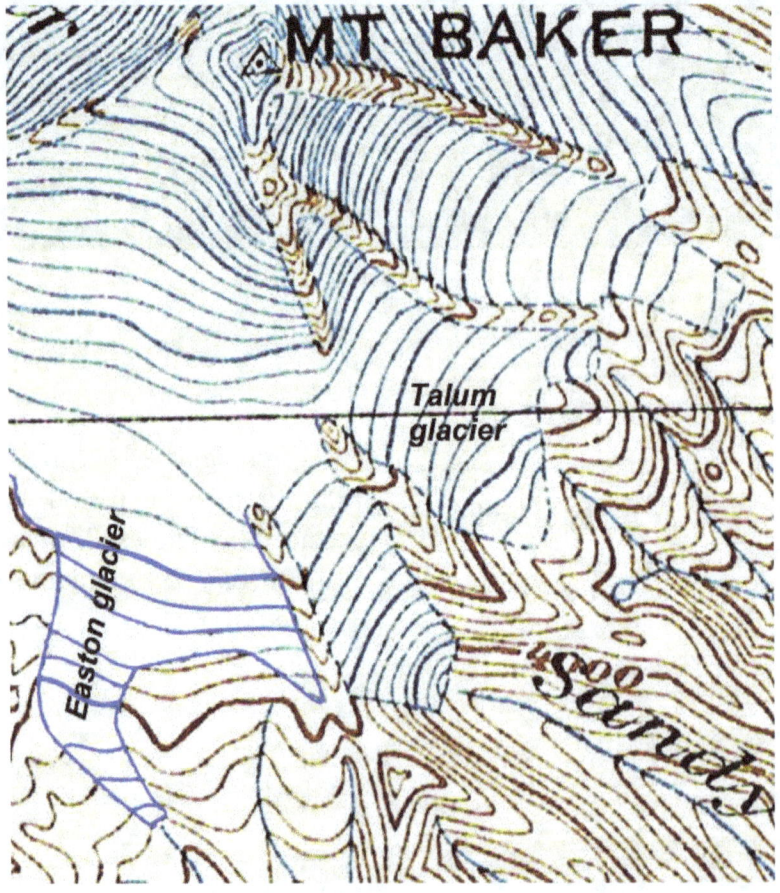

Figure 297. Talum glacier, 1915. (USGS map)

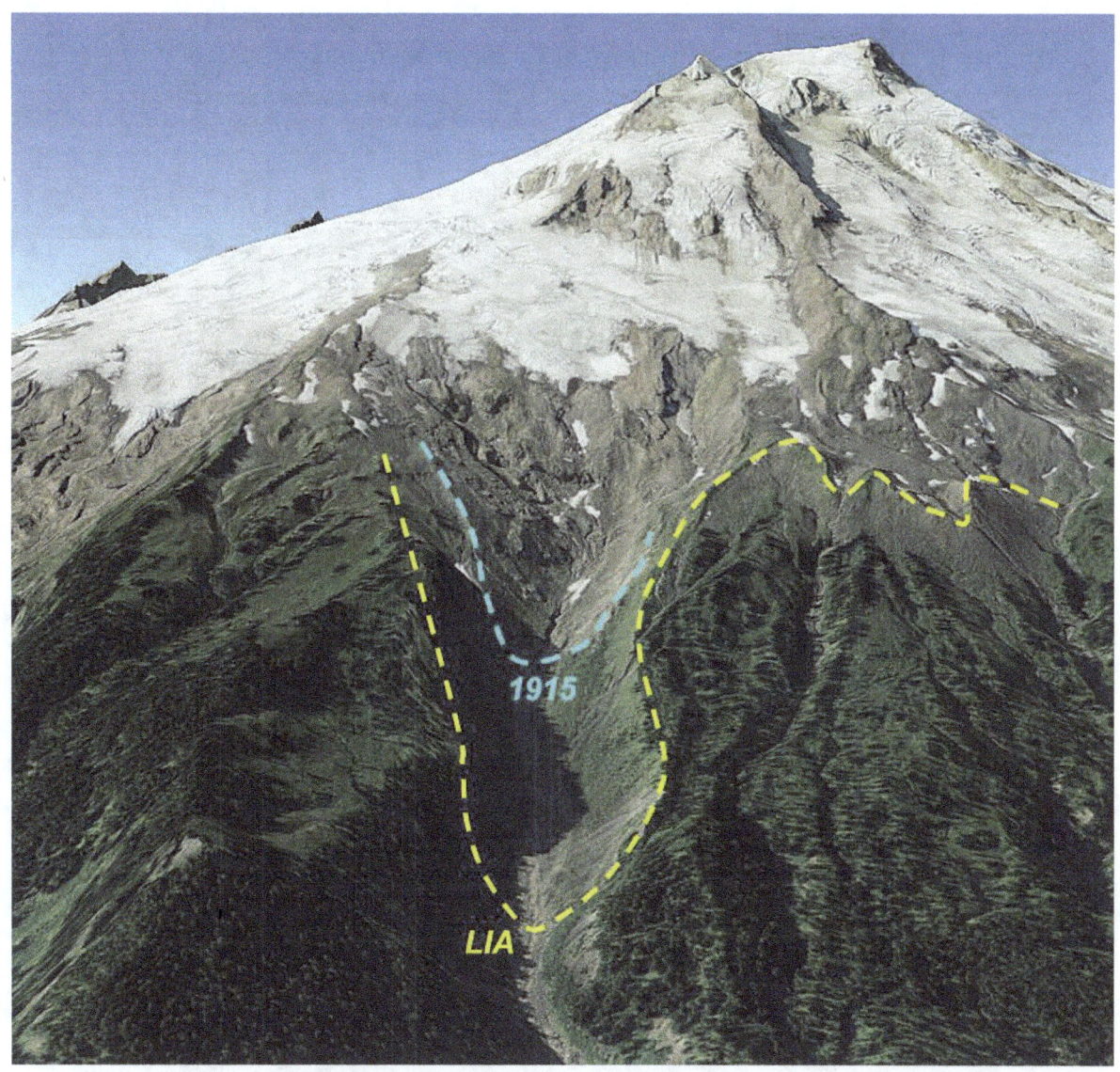

Figure 298. Little Ice Age (~1700 AD) and 1915 margins of the Talum glacier.

Glacier fluctuations from 1915 to present

Fluctuations of the Talum glacier from 1915 to present have not been studied in detail, but undoubtedly followed the same pattern as other glaciers on Mt. Baker, i.e., strong retreat from 1915 to about 1945-52, readvance from ~1956 to ~1980, and recession from ~1980 to ~2000. The strongest advance occurred from 1880 to 1915. Figures 299 and 300 show the position of the terminus in 1940 and figures 301–304 show the terminus in 1958, 1970, 1979, and 1987.

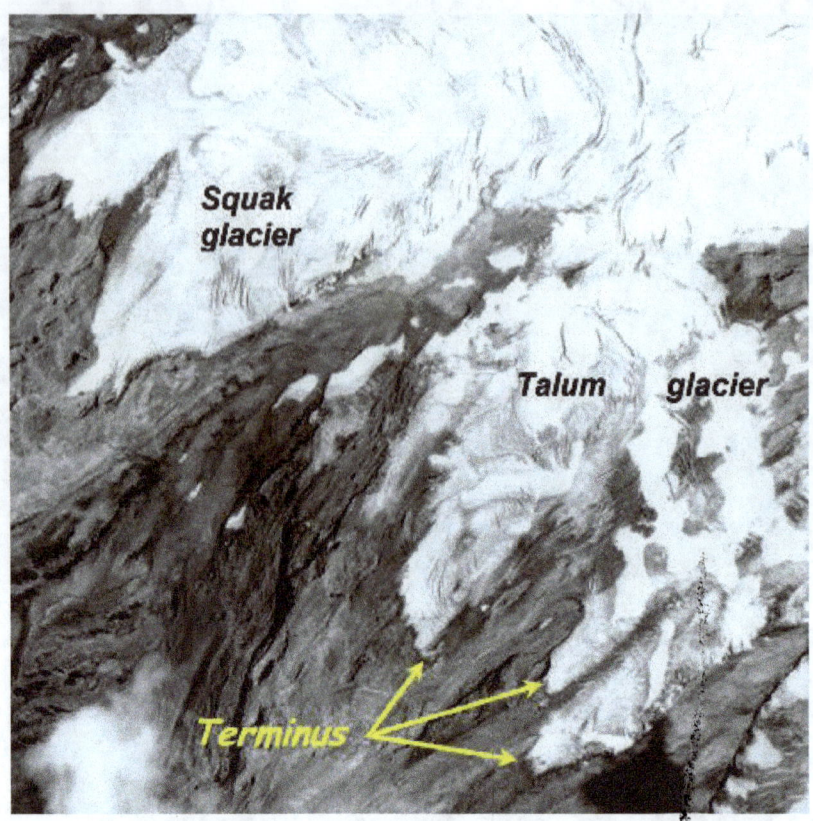

Figure 299. Talum glacier 1940.

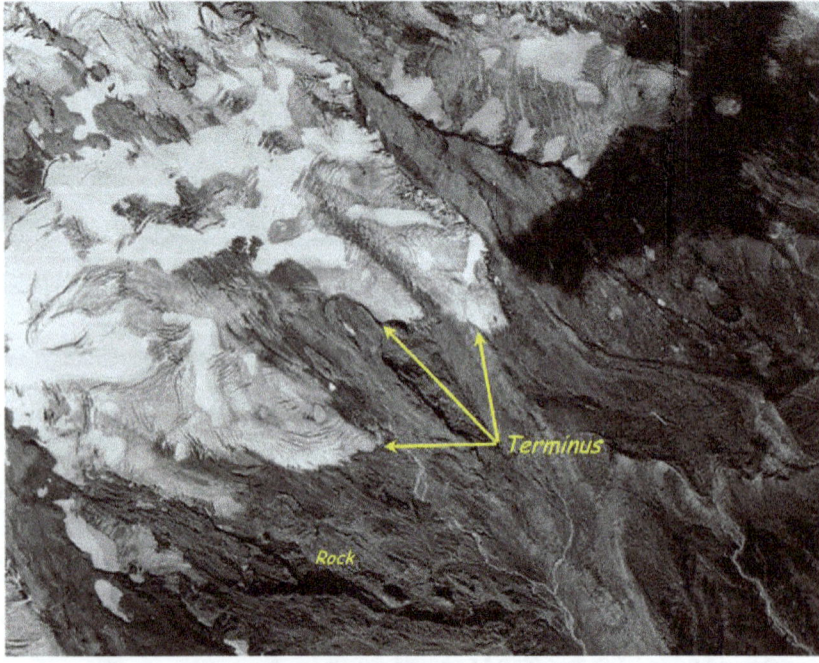

Figure 300. Talum glacier, 1940.

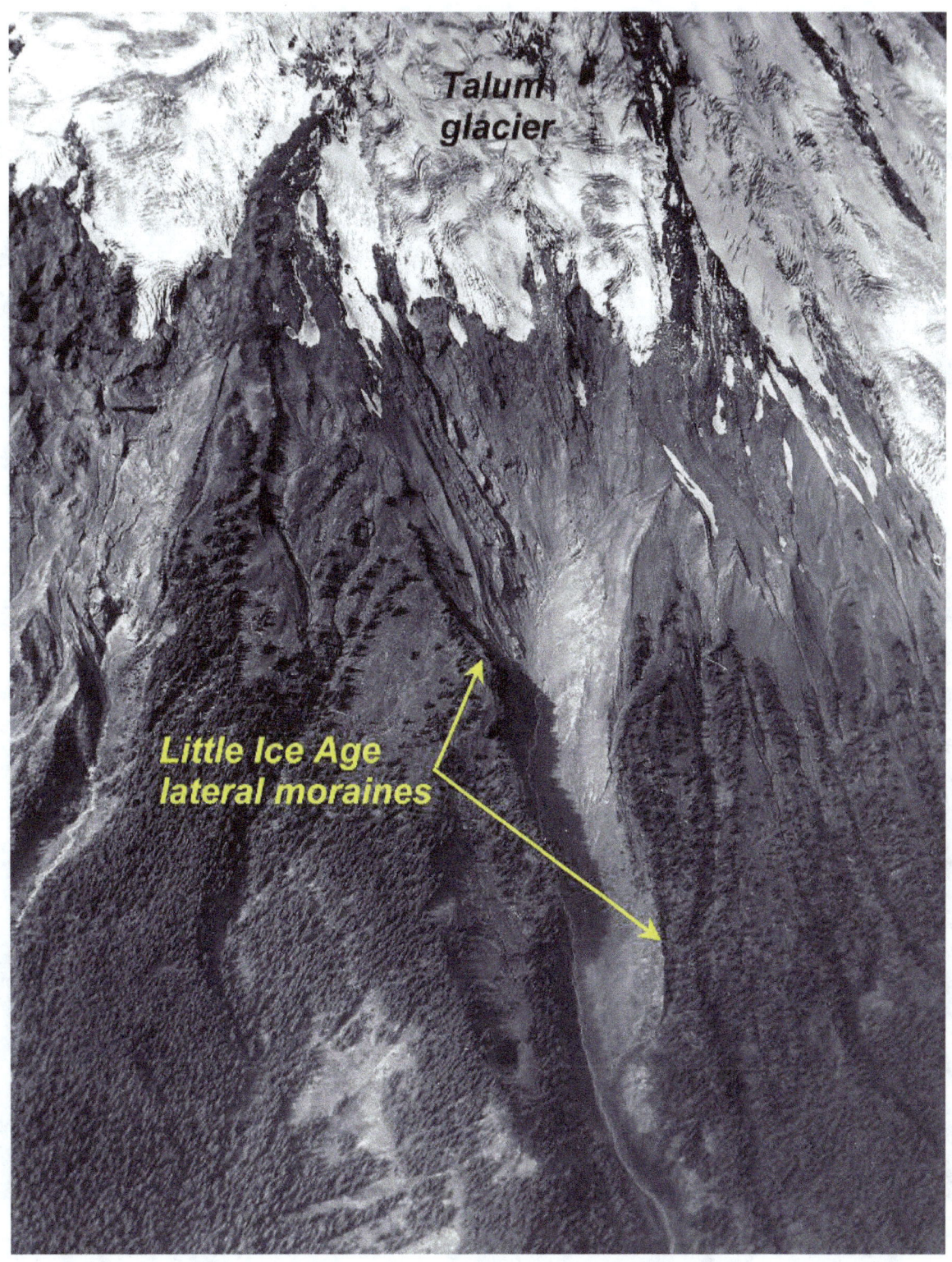

Figure 301. Talum glacier, 1958. (Photo by Austin Post)

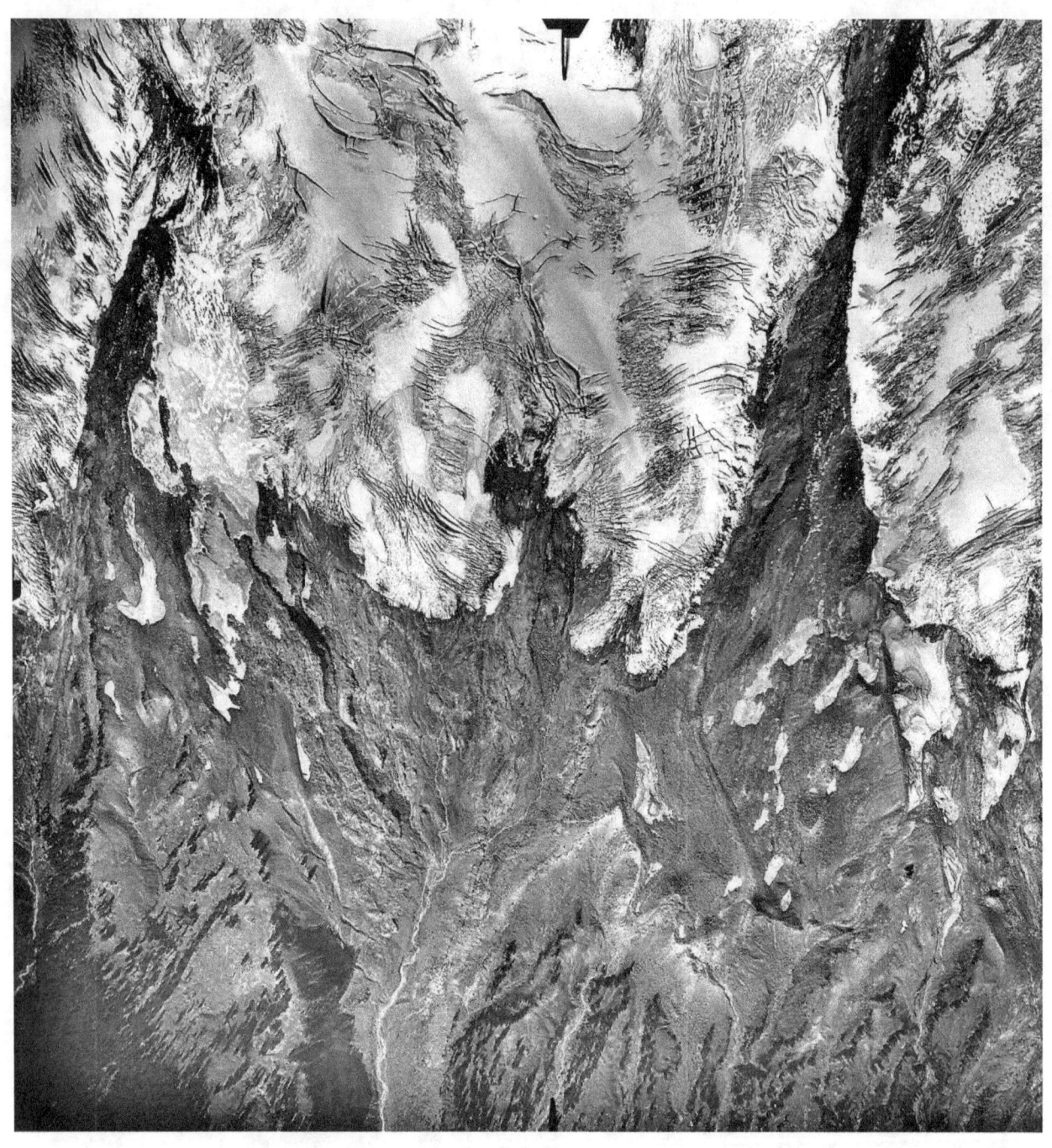

Figure 302. Talum glacier, 1970. (Photo by Austin Post).

Figure 303. Talum glacier, 1979. (Photo by Austin Post)

At the end of the ~1950 to 1979 cool period, the Talum glacier stopped advancing and began to retreat upvalley. Figures 304, and 305 show the terminus in 1987 and 2009. Comparison of the terminal position in 2009 and 2013 shows that the glacier appears to have stopped retreating sometime between 2009 and 2013.

Figure 304. Talum glacier, 1987. (Photo by Austin Post)

Figure 305. Talum glacier, 2009.

Figure 306 shows the position of the Talum glacier terminus in 1952 and Figure 307 shows the terminus in 2014. Comparison of these maps (Fig. 308) shows that the glacier terminus in 2014 was more than half a mile downvalley from its 1952 position.

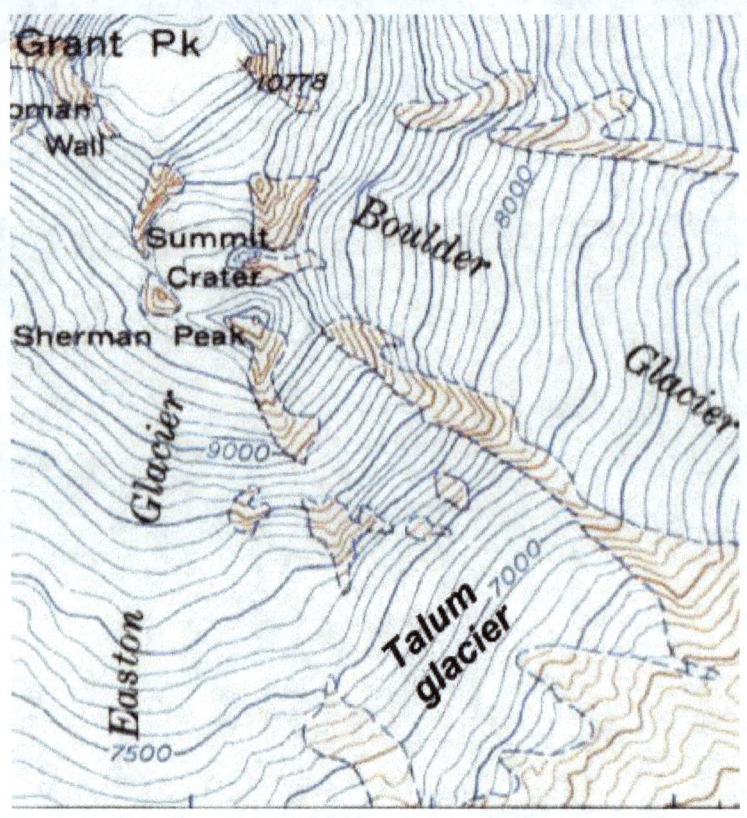

Figure 306. Talum glacier in 1952. (USGS map)

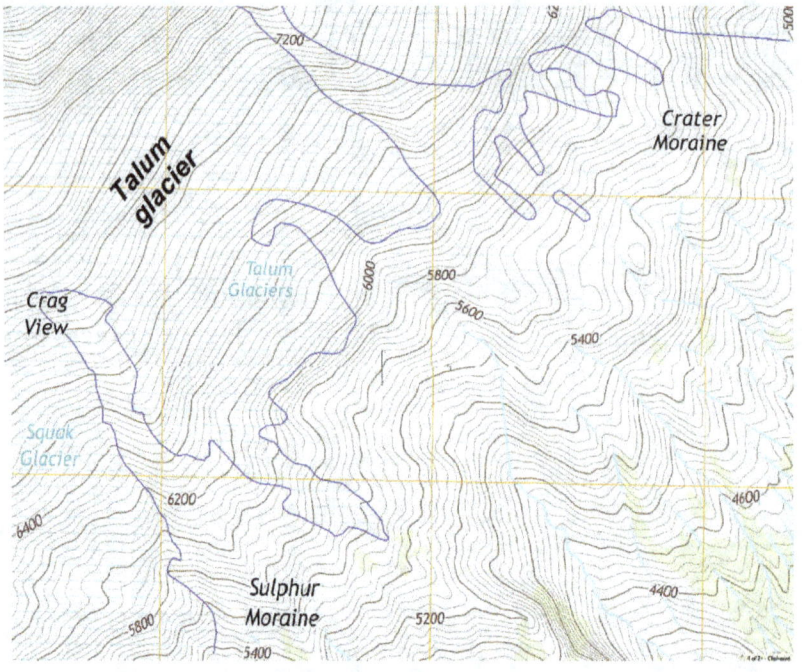

Figure 307. Talum glacier in 2014 (USGS map)

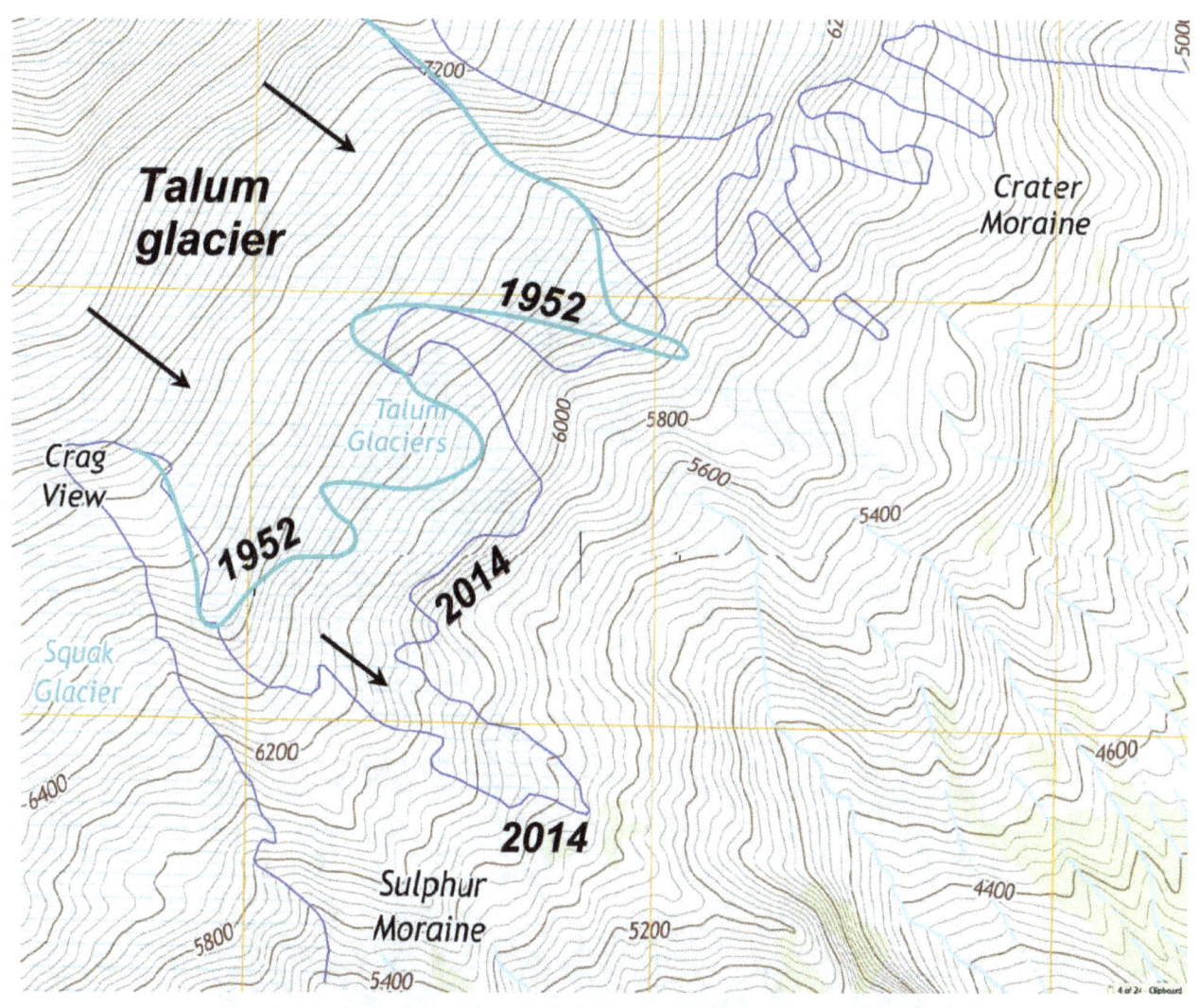

Figure 308. Comparison of 1952 and 2014 termini of the Talum glacier. (USGS map)

Boulder Glacier

The Boulder glacier flows from Sherman crater into the headwaters of Boulder Creek (Figs. 309-311).

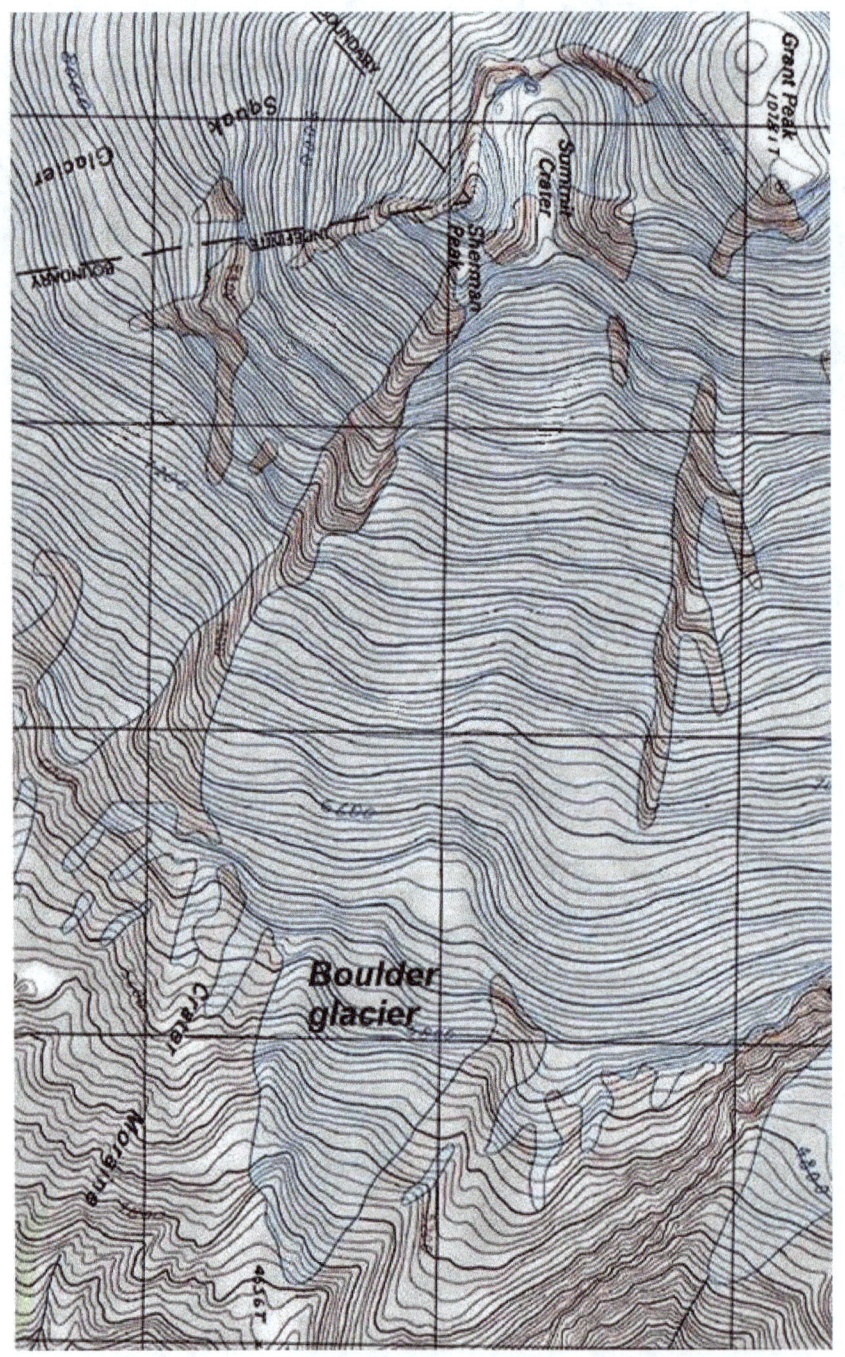

Figure 309. Topographic map of Boulder glacier.

Figure 310. Boulder glacier, 1987. (Photo by Austin Post)

Younger Dryas and Little Ice Age moraines

High lateral moraines along the valley sides below the Boulder glacier (Fig. 312) probably date to the Younger Dryas and Little Ice Age, but they have not been dated. Much of the Boulder Creek valley is filled with unconsolidated sediments interbedded with lava flows. Early Holocene (8,700 ^{14}C yrs)) sediments are exposed along the valley walls of Boulder Creek about midway between the present glacier and Baker Lake (Fig. 313). A thick, massive poorly sorted lower unit that is either a volcanic mudflow or glacial till contains wood that has been dated at 8,700 ^{14}Cyrs. It is overlain by lava and a volcanic ash, which apparently was erupted from Mt. Baker, (Fig. 313). Another massive, poorly sorted layer contains two volcanic ashes. The composition of the lower ash suggests that it may have come from Mt. Rainer and the upper ash is Mazama ash (6,800 ^{14}C yrs old). At the top of the section is another Mt. Baker lava flow.

Figure 311. Boulder glacier. The bare ground beyond the glacier terminus was occupied by ice about 1915.

Figure 312. Boulder glacier and forest trim lines at the 1920 glacier terminus.

Figure 313. [left] Sediments and lava in the Boulder Creek valley. Unconsolidated sediments filling the Boulder Creek valley.

During the Little Ice Age, about 500 years ago, the Boulder glacier extended about one mile downvalley from the present terminus (Fig. 314).

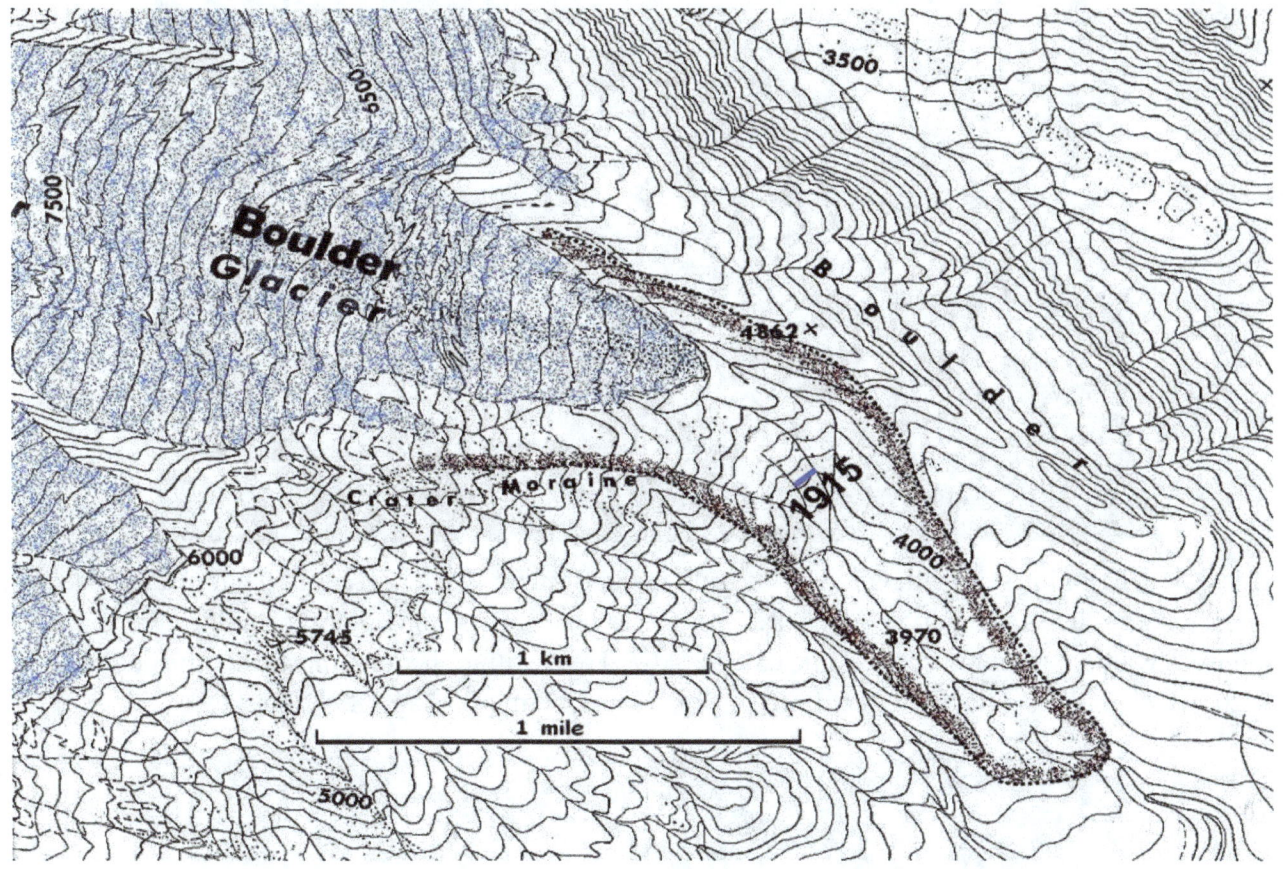

Figure 314. Little Ice Age extent of the Boulder glacier (brown). (USGS map)

Several conspicuous trim lines in the forest in the Boulder Creek valley below the present glacier can be used to date the retreat of the glacier (Figs. 315-317) during the Little Ice Age. Tree rings of the oldest trees unaffected by the glacier date back to 1558 AD. This date is probably close to the age of the Little Ice Age maximum of the Boulder glacier. Successive forest trimlines upvalley marking positions of former glacier termini date to about 1838, 1867, and 1920. In 1915, the glacier terminus was well downvalley from the present terminus (Fig. 314), but upvalley from the Little Ice Age maximum.

Figure 315. Trim lines of Little Ice Age margins of the Boulder glacier. Trees in the forest outside the outermost trim line date to 1558 AD. Cores from trees inside the second oldest trim line date to 1838 and trees within the next inner trim line date to 1867. Trees within the innermost trim line date to 1920, which dates the glacier retreat from the 1915 maximum. (Tree ages from Burke, 1972)

Figure 316. Trim line from the 1915 position of the Boulder glacier.

A 1908 photo of the glacier (Fig. 317) shows thick ice well downvalley from the present terminus.

Figure 317. Thick ice of the Boulder glacier in 1908. The valley here is now ice-free.
(Photo by Asahel Curtis)

The 1909 terminus of the Boulder glacier is shown on Figure 318 and the 1915 terminus is shown on Figure 319. This terminal position marks the maximum extent of the Boulder glacier in this century as a result of the 1880 to 1915 cool period.

Figure 318. Boulder glacier on 1909 USGS map.

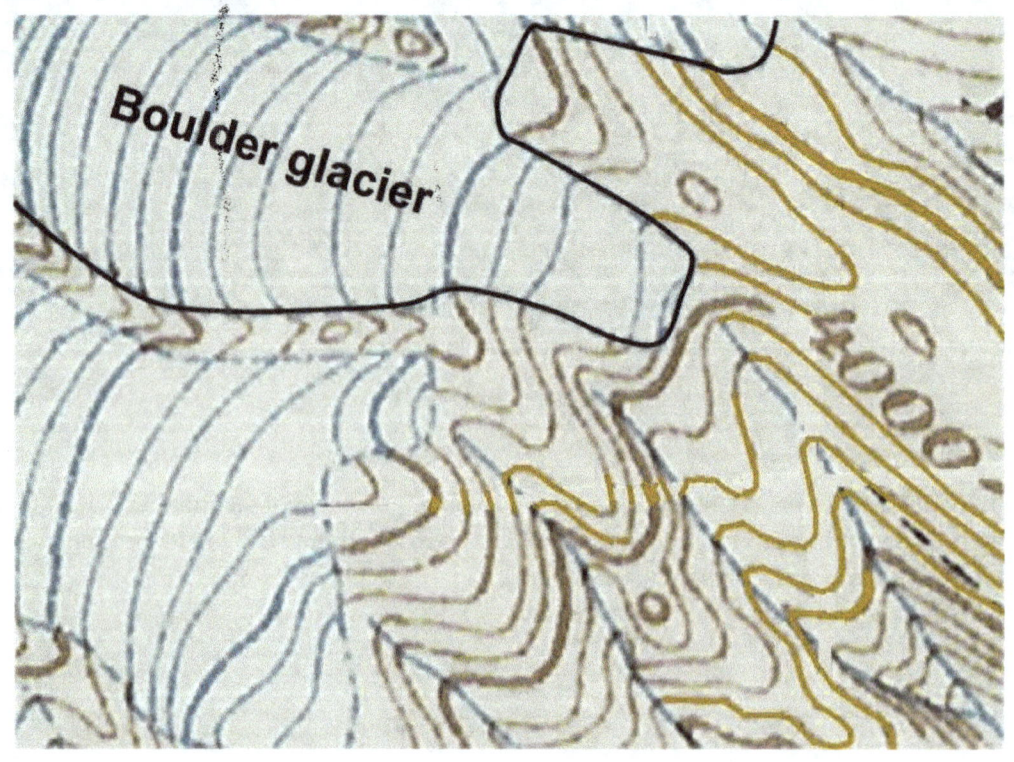

Figure 319. Boulder glacier on 1915 USGS map.

Glacier fluctuations since 1915

Direct observation of fluctuations of the Boulder glacier began in 1931 and continued to 1957. Glacial changes are also shown on air photos beginning in 1940. The Boulder glacier retreated upvalley from its 1915 position as a result of the 1915 to 1945 warm period.

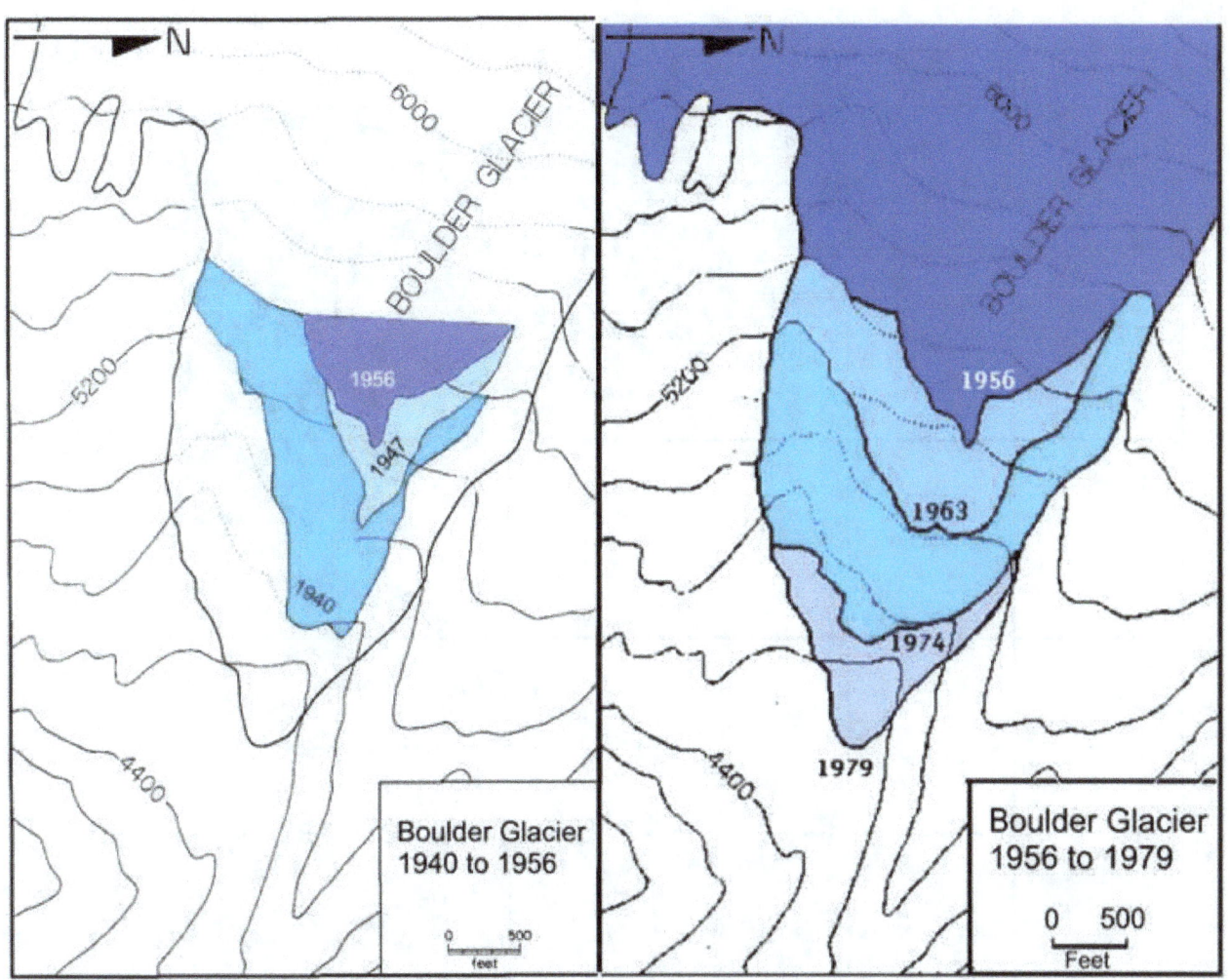

Figure 320. Advance and retreat of the Boulder glacier from 1940 to 1990. (Modified from Harper, 1992).

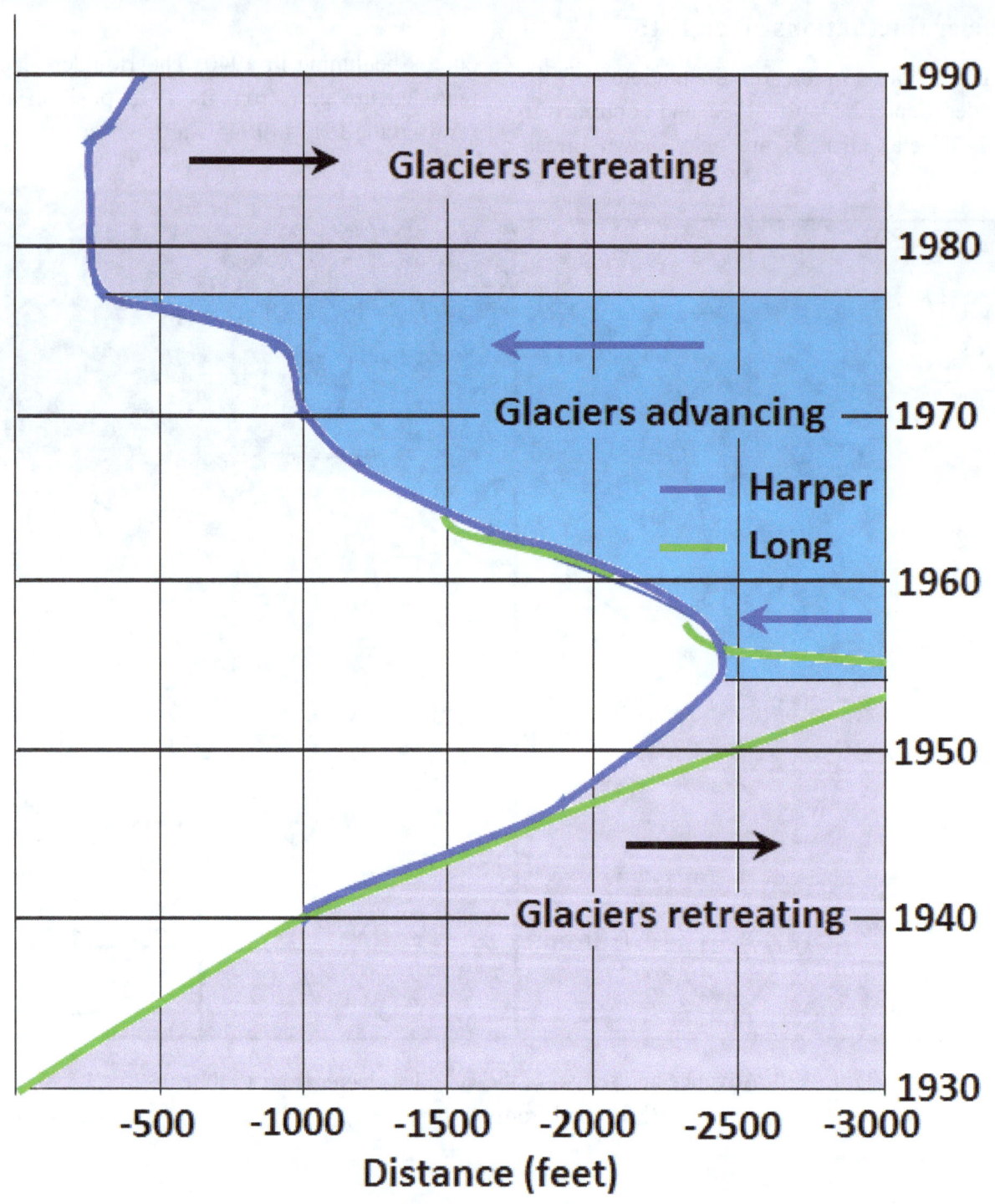

Figure 321. Graph showing the amount of advance and retreat of the Boulder glacier, 1931 to 1990.

The glacier retreated 1,000 feet from 1931 to 1940 and 2,150 feet from 1940 to 1954. The total retreat from 1931 to 1954 was 3,150 feet. Air photos show the glacier receded 900 feet from 1940 to 1947 (Fig. 323) and 551 feet from 1947 to 1956, a total of 1,447 feet from 1940 to 1956.

The retreat halted in 1954 and the glacier began to advance. The terminus advanced 203 feet from 1954 to 1955 and 630 feet from 1955 to 1957 (Figs. 324, 325), a total of 833 feet from 1954 to 1957.

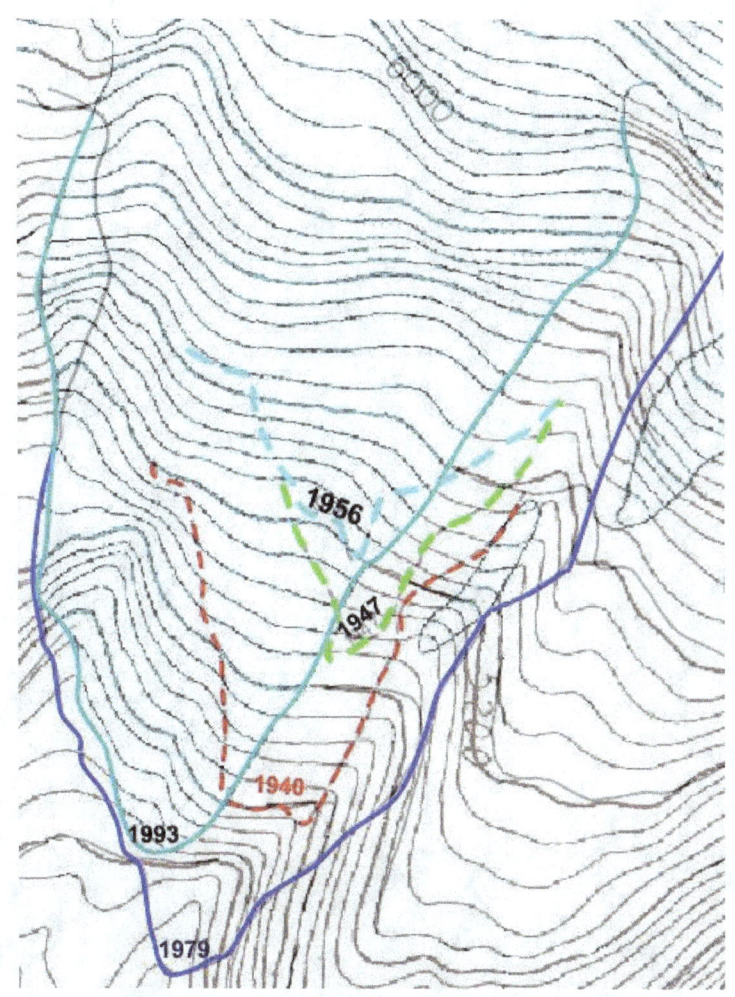

Figure 322. Positions of the Boulder glacier terminus from 1940 to 1979.

Figure 323. Boulder glacier, 1947.

Figure 324. The Boulder glacier in 1955. (Photo by Austin Post)

Figure 325. The Boulder glacier, 1955. (Photo by Austin Post)

Figure 326. Boulder glacier, 1969. (Photo by Austin Post)

Air photos show the glacier advanced 804 feet from 1956 to 1963 and 456 feet from 1963 to 1967, a total of 1,260 feet from 1956 to 1967. The glacier was observed to advance 834 feet from 1964 to 1969 (Figs. 327, 328). Air photos show the terminus advanced 456 feet from 1963 to 1967.

Figure 327. Terminus of the Boulder glacier in the late 1960s during the strong 1950-1980 readvance. Note the steepness of the ice front, typical of advancing glaciers.

From 1967 to 1974 (Figs. 326–328), the glacier terminus advanced 292 feet after which the advance accelerated considerably, advancing 590 feet from 1974 to 1977. The advance slowed to only 33 feet from 1977 to 1979 and only 16 feet from 1979 to 1986. Other Mt. Baker glaciers generally reached their maximum recent extent about 1979 when the climate changed from cool to warm, but the Boulder glacier was rather sluggish in its response, continuing to advance for another seven years. After 1986, the Boulder terminus began to retreat once more.

Figure 328. Boulder glacier, 1974.

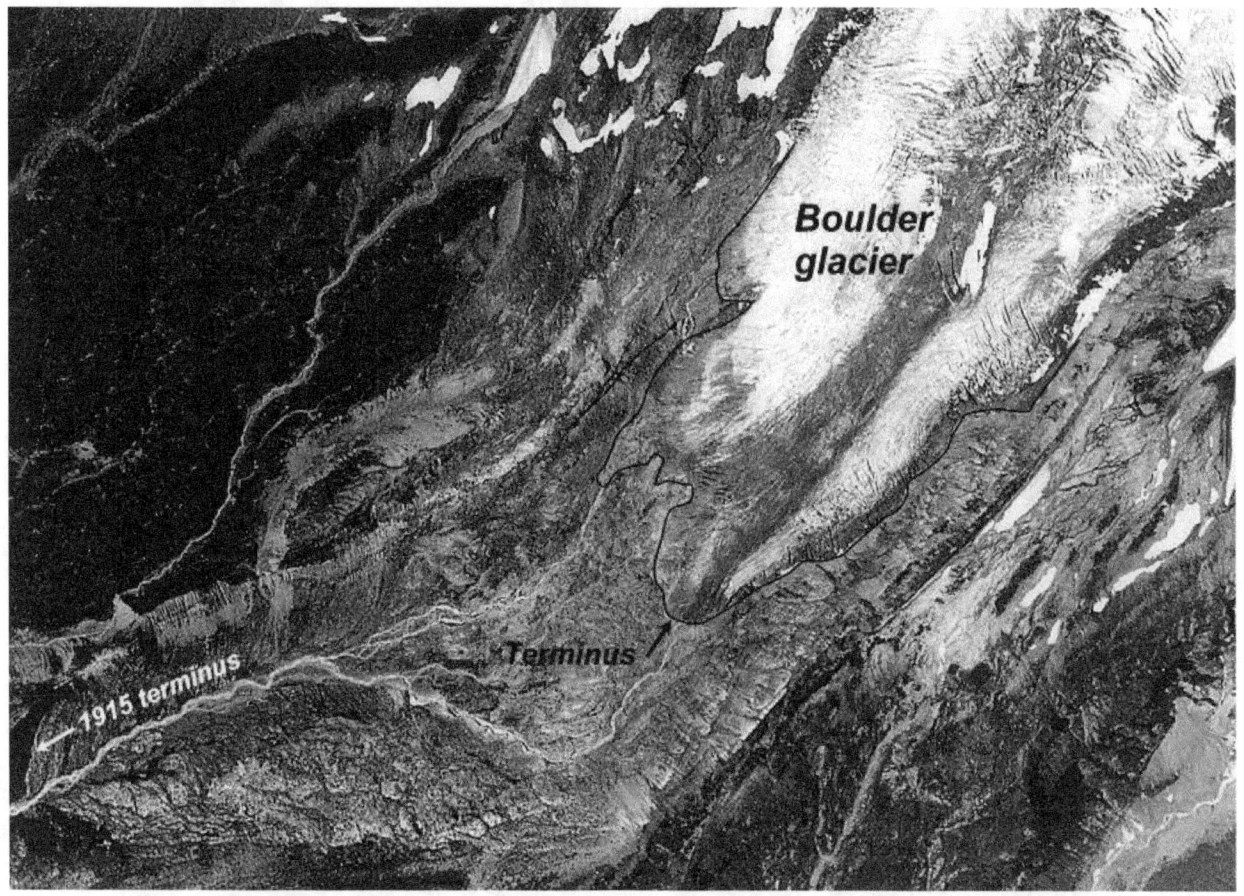

Figure 329. Boulder glacier, 1985. (Photos by Austin Post)

The Boulder glacier retreated from the early 1980s until about 2009 (Figs. 329–330). Figures 331 and 332 show the terminus of glacier in 2002 and 2009. Comparison of photos of the 2009 terminus with photos of the terminus in 2013 shows the glacier spilling over a ledge and a rock outcrop at part of the ledge that was ice–covered in 2009, but the terminus appears to be in the same place. 2015 photos show no retreat of the terminus from its 2013 position. Thus, the glacier appears to have stopped retreating.

Figure 330. Boulder glacier, 1993. (Photo by Austin Post)

Figure 331. Boulder glacier, 2002. (Photo by Austin Post)

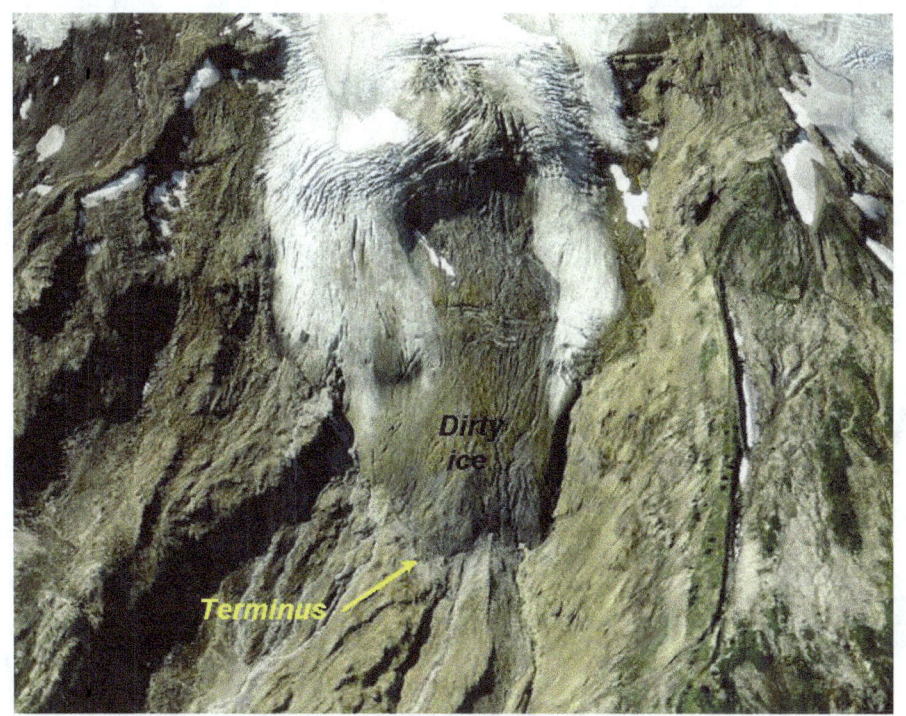

Figure 332. Boulder glacier, 2009.

Figure 333. Streamlined bedrock sculpted by overriding by the Boulder glacier.

Comparison of the Boulder glacier terminus in 1952 (Fig. 334) and 2014 (Fig. 335) shows that the 2014 terminus was more than half a mile downvalley from the 1952 terminus (Fig.335).

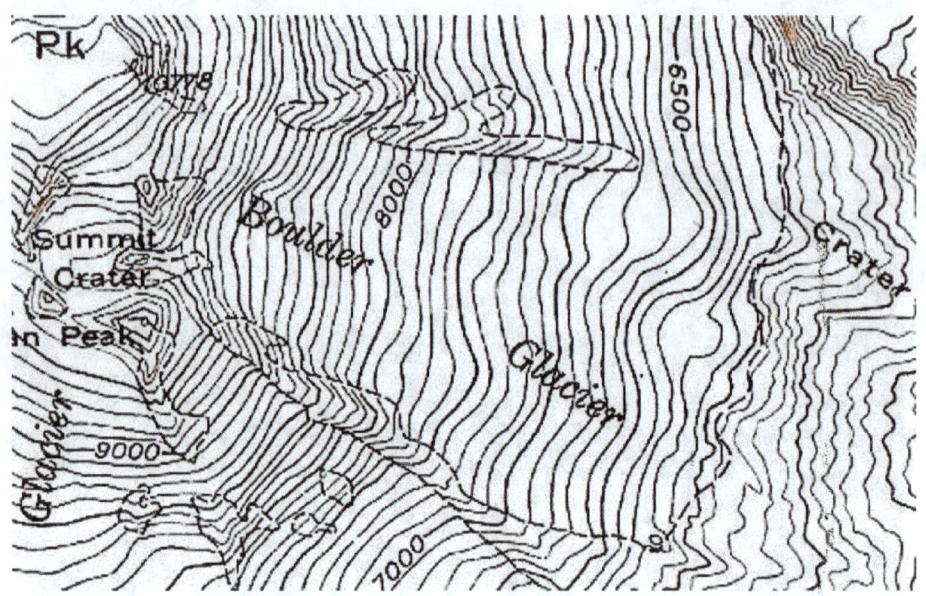

Figure 334. 1952 map of the Boulder glacier. (USGS map)

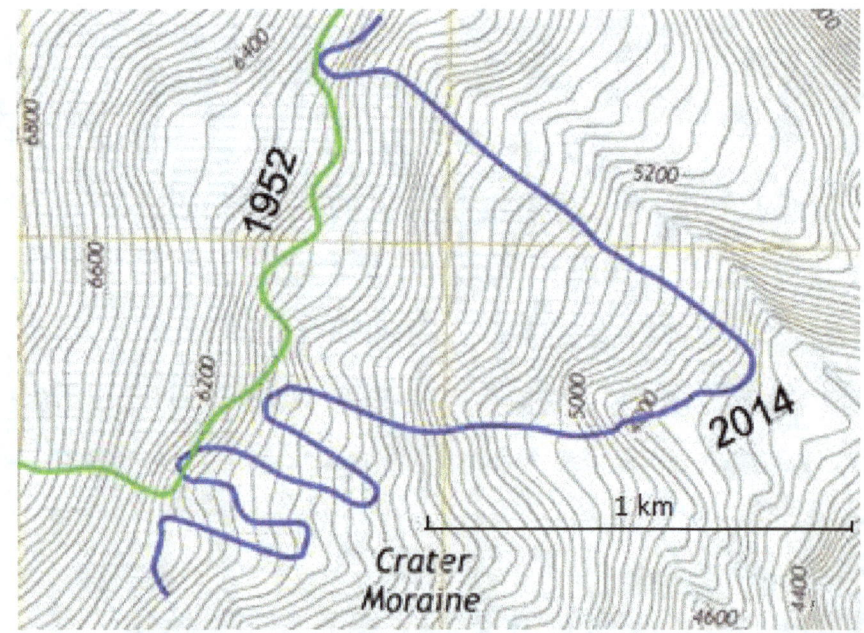

Figure 335. Boulder glacier terminus in 2014 (blue) and 1952 (green). The 2014 terminus is more than half a mile downvalley from the 1952 position. (USGS map)

Park glacier

The Park glacier begins on the NE flank of Grant Peak, Mt. Baker's summit cone (Figs. 336, 337). About halfway down, it makes a right angle bend and drops into a deep trough between Park Cliffs and Lava Divide. Avalanches from a broad expanse of ice due east of the Grant Peak cascade over Park Cliffs and feed the lower part of the glacier. The valley sides along Park Cliffs and upper Park Creek are so steep that lateral moraines are not preserved.

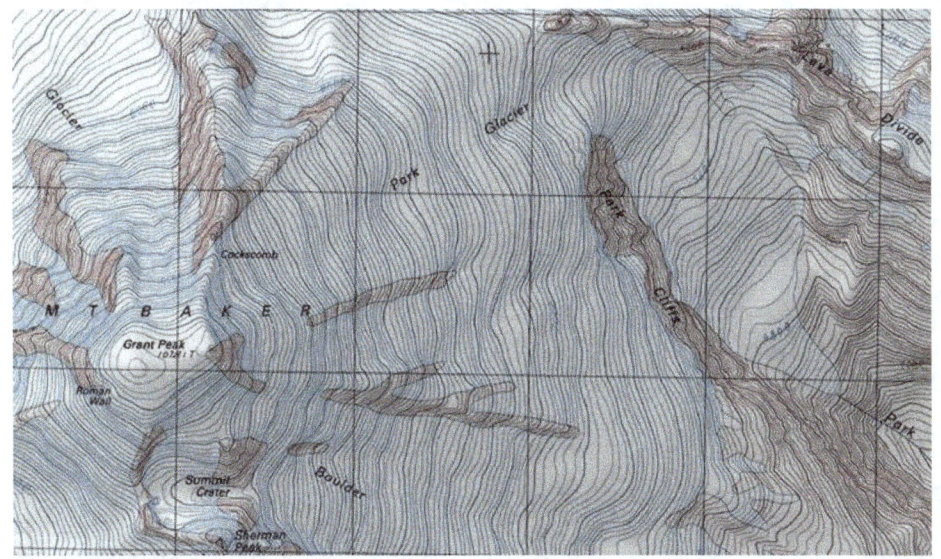

Figure 336. Park Glacier.

Figure 337. Park glacier, 1988.

Little Ice Age

The extent of the Park glacier during the Little Ice Age (~1500 AD) is not well known, but lateral moraines downvalley from the present glacier terminus (Fig. 341) are thought to have been deposited then.

Figure 338. Park glacier moraines (brown dashed line) of the Little Ice Age. (USGS map)

1915 to present

The Park glacier reached its maximum extent of the past century in 1915 at the end of the 1880–1915 cool period. The terminus of the glacier is shown on the 1909 (Fig. 339) and 1915 (Fig. 340) USGS topographic maps. Figure 341 shows the position of the terminus about 1915. From 1915 to about 1950, the Park glacier retreated upvalley, reaching its highest position in the late 1940s (Fig. 342) or early 1950s.

The Park glacier advanced strongly from the early 1950s to 1979 (Figs. 343–345). From 1964 to 1974, the glacier advanced 570 feet. In 1977, the eastern Pacific Ocean flipped from its cool mode to its warm mode and the Park glacier began to retreat once again (Figs. 346–348).

Figure 339. Park glacier. 1909. (USGS map)

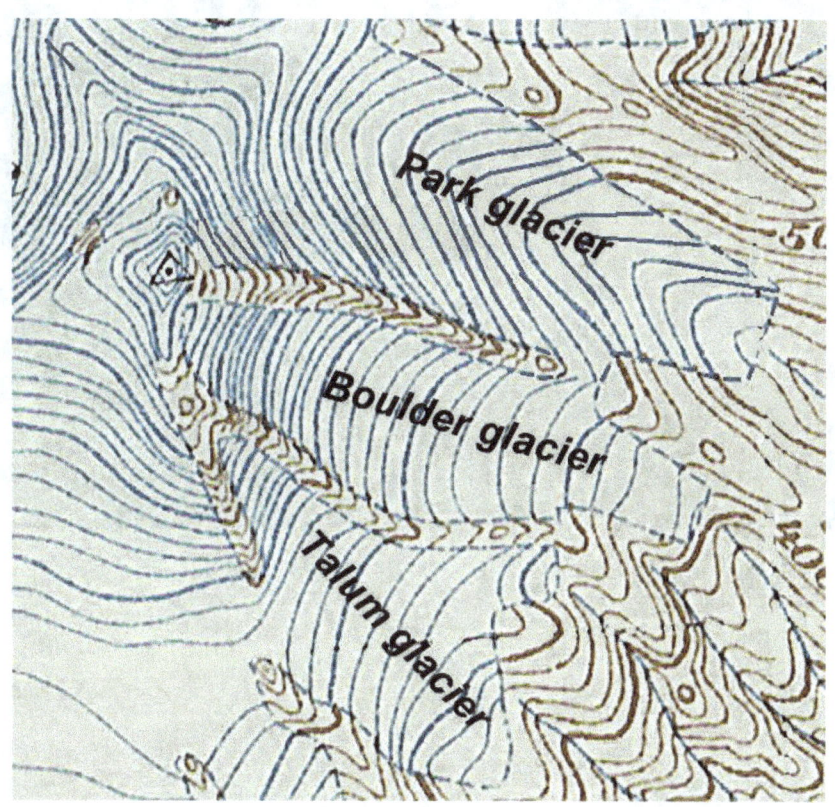

Figure 340. Park glacier, 1915. (USGS map)

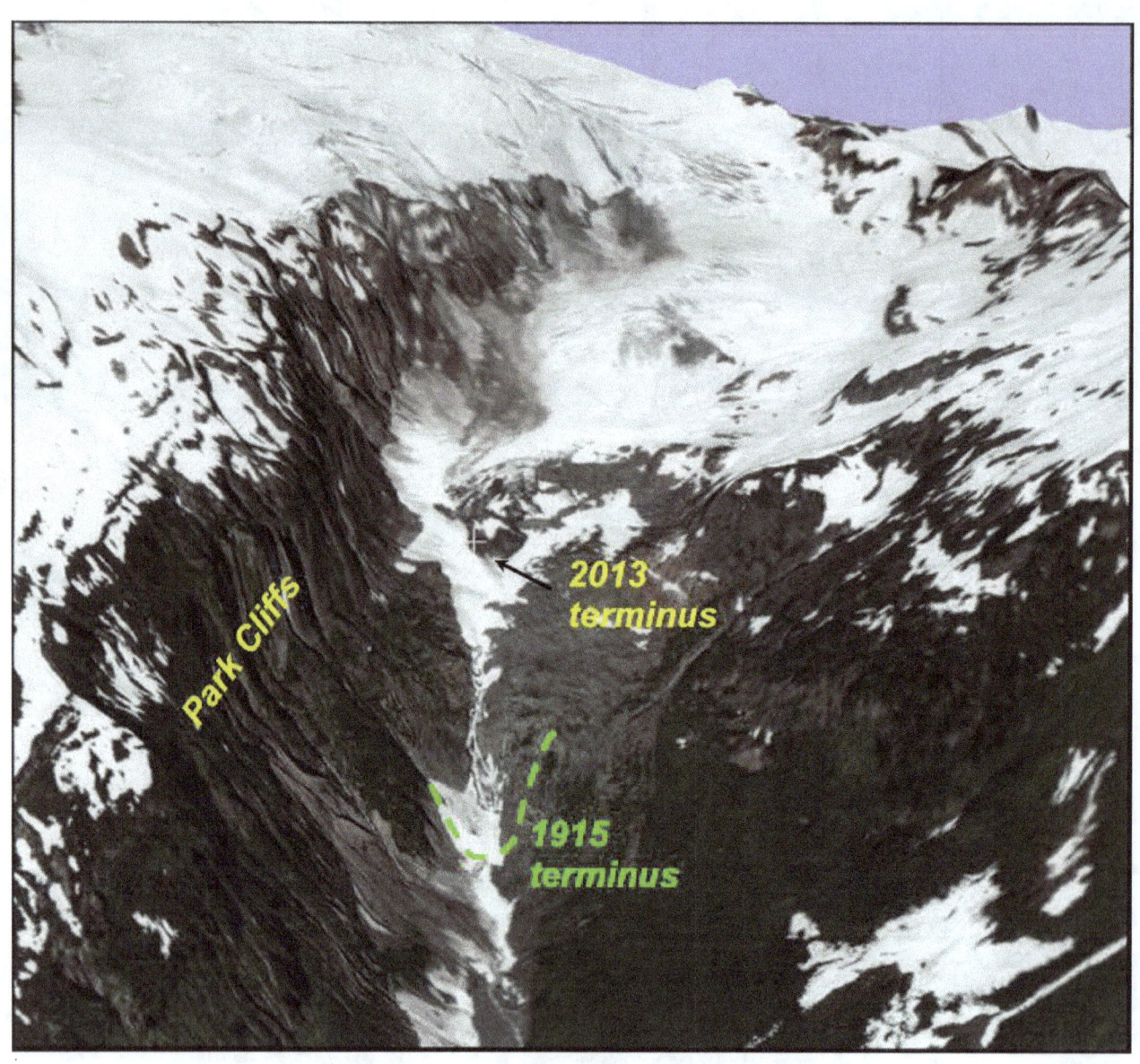

Figure 341. 1915 terminus, Park glacier.

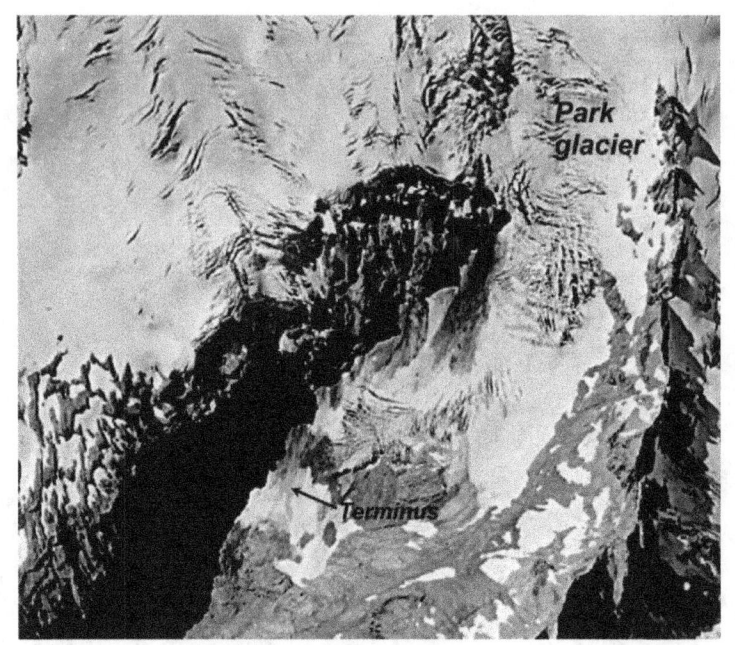

Figure 342. Park glacier, 1947.

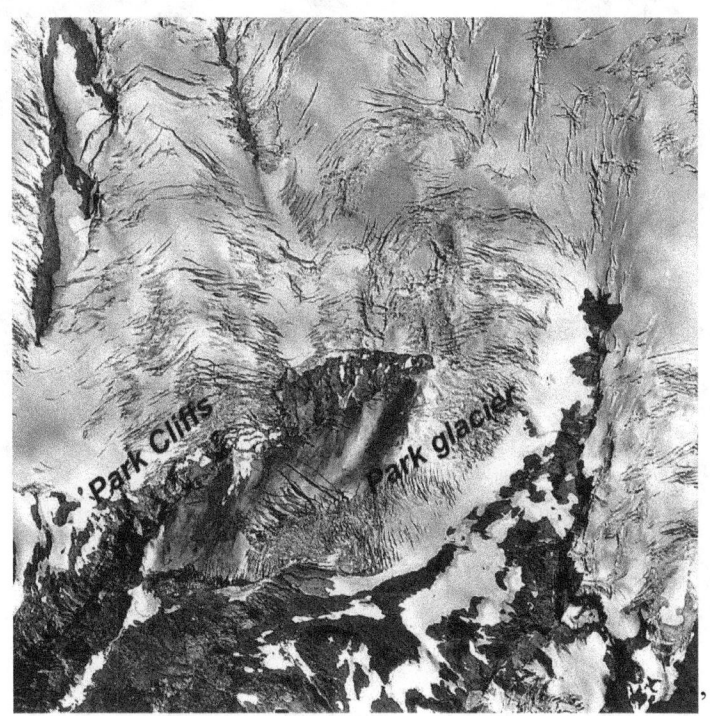

Figure 343. Park glacier, 1956. (Photo by Austin Post)

Figure 344. Park glacier, 1962. (Photo by Austin Post)

Figure 345. Park glacier, 1975. (Photo by Austin Post)

Figure 346. Park glacier, 1987. (Photo by Austin Post)

Figure 347. Park glacier, 2002. (Photo by Austin Post)

Figure 348. Park glacier, 2009.

Comparison of the Park glacier terminus in 1952 (Fig. 349) and 2014 (Fig. 350) shows that the 2014 terminus was more than half a mile downvalley from the 1952 terminus.

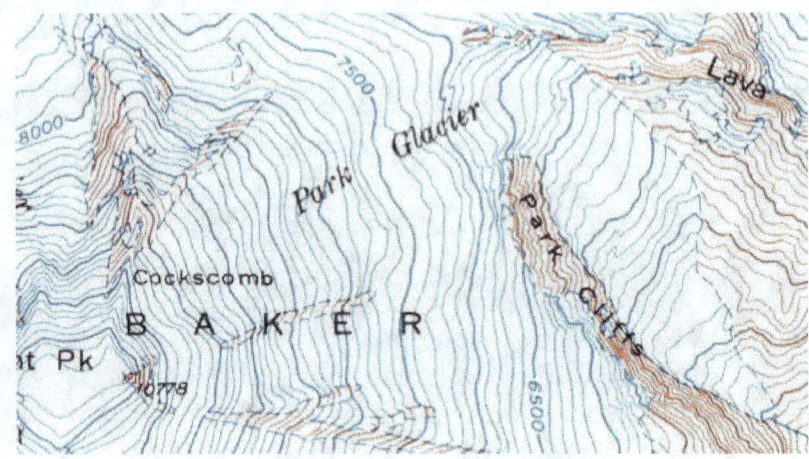

Figure 349. Park glacier in 1952 (USGS map)

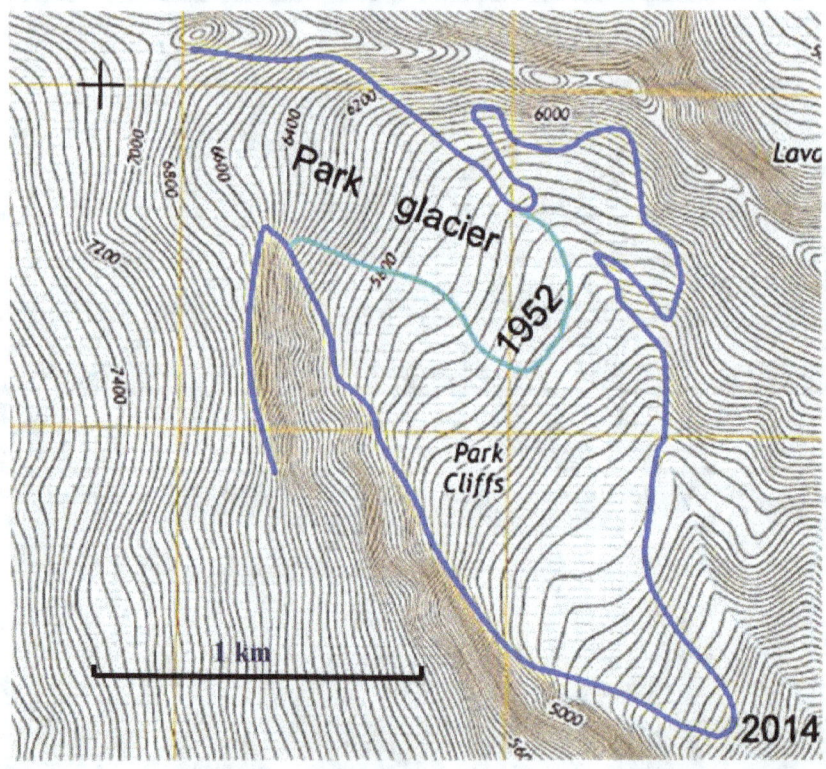

Figure 350. Comparison of the Park glacier in 2014 (blue line) and 1952 (green line). The 2014 terminus is more than half a mile downvalley from the 1952 position. (Modified from USGS map)

Rainbow glacier

The Rainbow glacier flows down the NE flank of Mt. Baker into Avalanche Gorge in the headwaters of Rainbow Creek (Figs. 351-353). The upper part of the glacier is contiguous with the Mazama and Park glaciers.

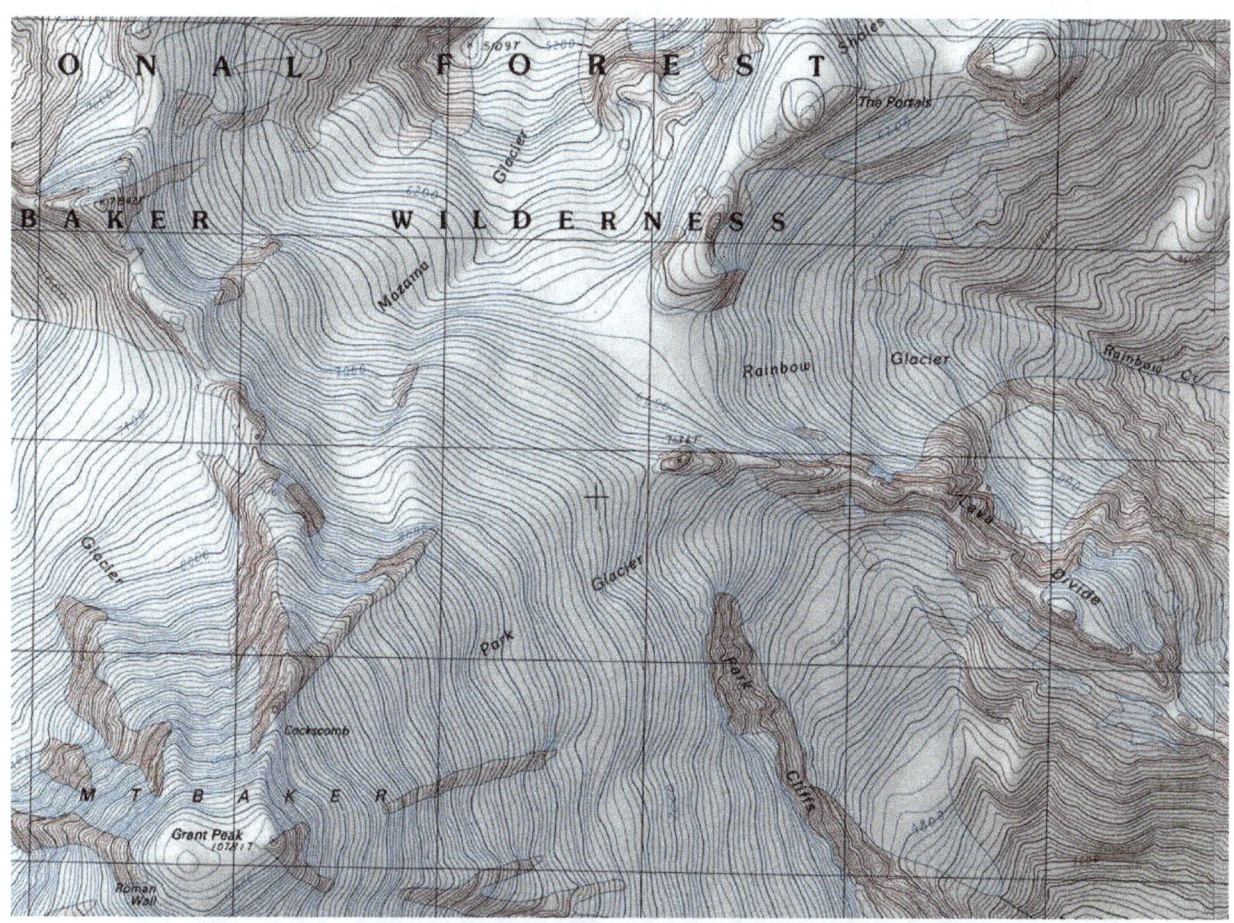

Figure 351. Rainbow glacier.

Figure 352. Rainbow glacier.

Figure 353. Rainbow glacier.

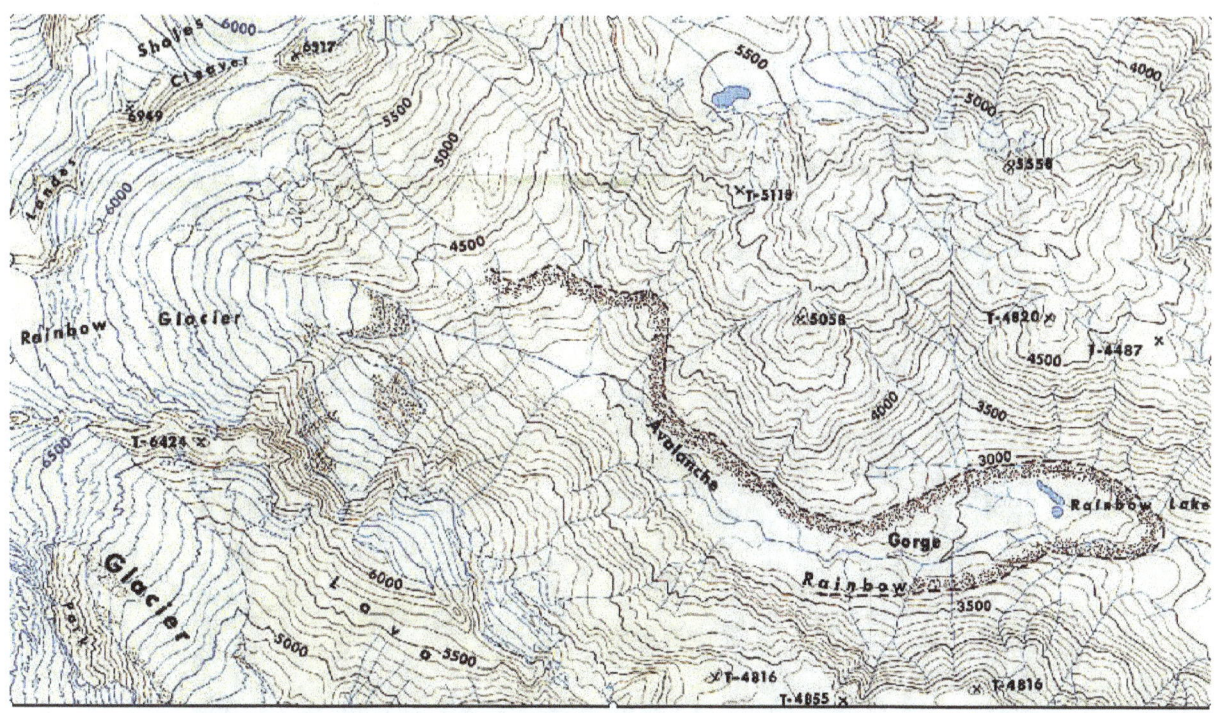

Figure 354. Little Ice Age moraine of the Rainbow glacier. (Modified from USGS map)

Figure 355. Rainbow Little Ice Age moraine (dashed yellow line).

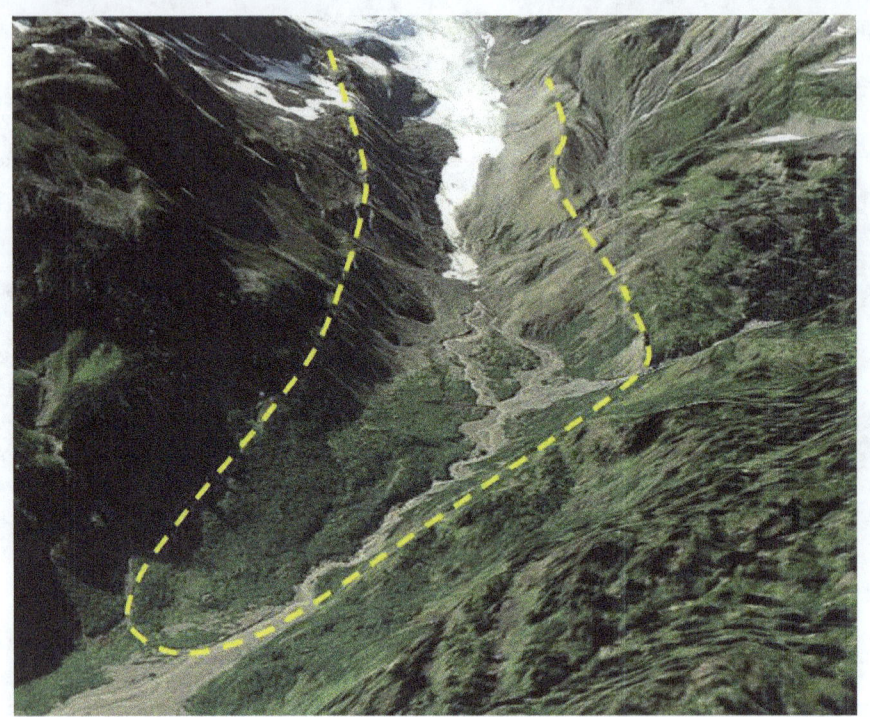

Figure 356. 1915 (?) moraines of the Rainbow glacier.

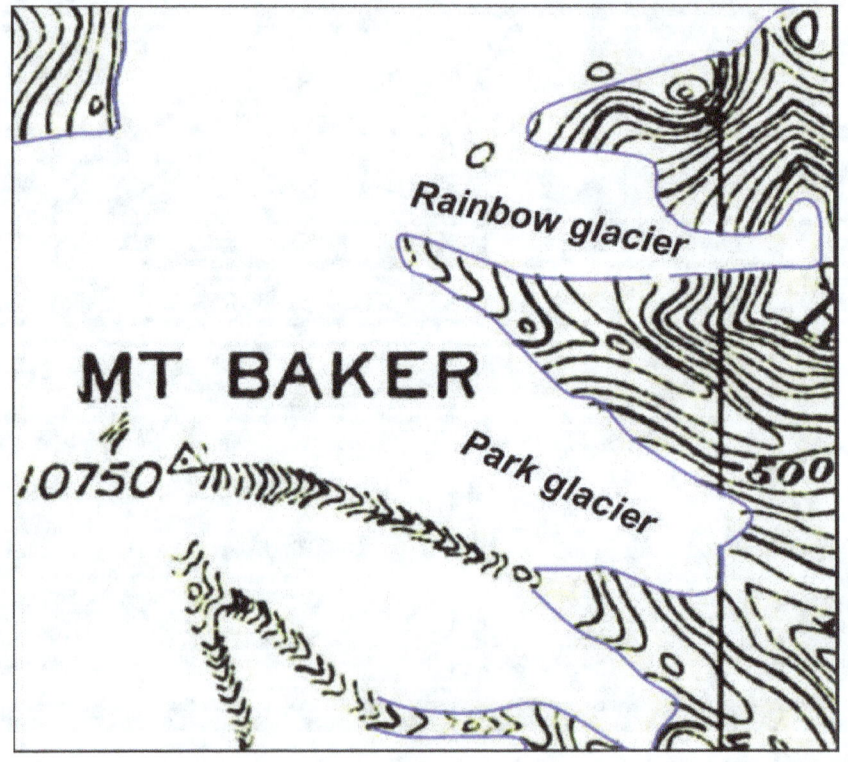

Figure 357. Rainbow glacier, 1909. (USGS map)

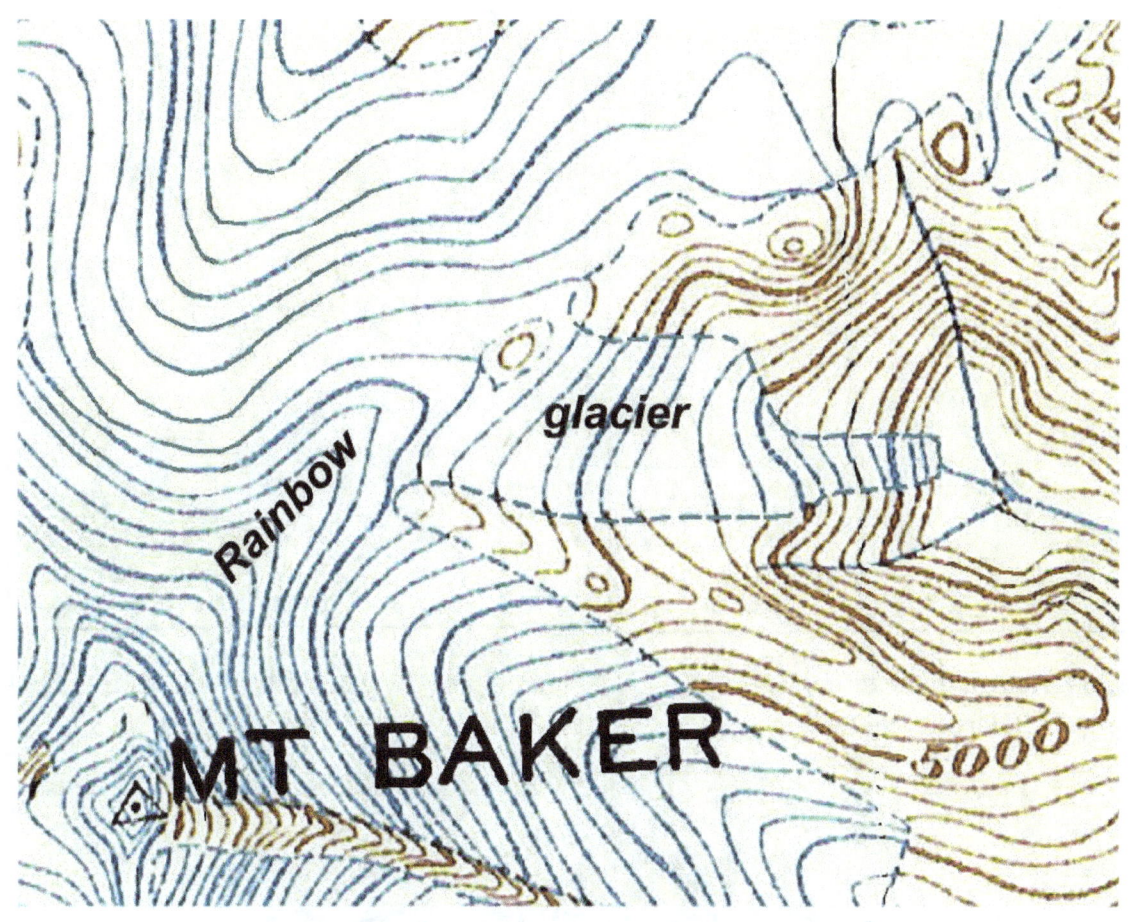

Figure 358. Rainbow glacier, 1915. (USGS map)

About 1945, the global climate changed from warm to cool and by 1947, recession of the Rainbow glacier ended and the glacier began to advance. The terminus advanced 30 feet from 1947 to 1956, 387 feet from 1956 to 1963, and 318 feet from 1963 to 1967. From 1967 to 1974, the terminus advanced 562 feet and between 1967 and 1979, it advanced 917 feet.

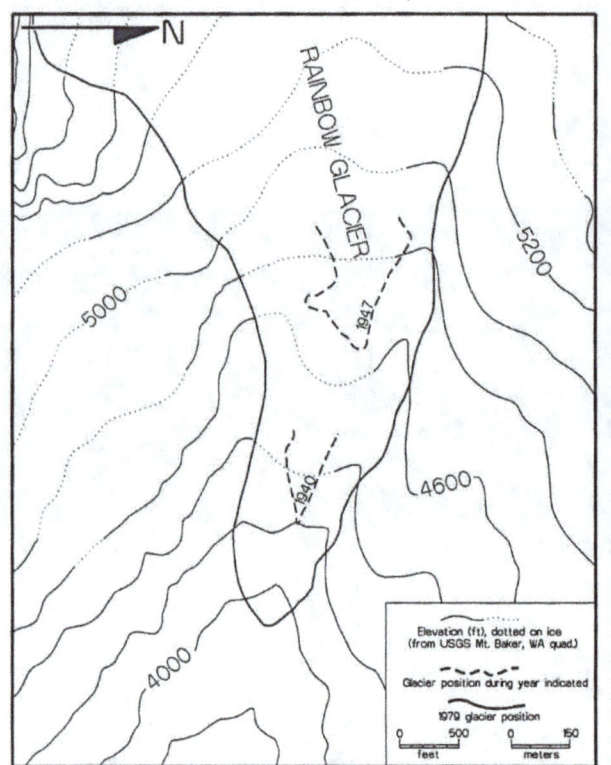

Figure 359. Retreat of the terminus from 1940 to 1977. (Harper, 1992)

Figure 360. Advance of the terminus from 1947 to 1947. (Harper, 1992)

Figure 361. Rainbow glacier 1940.

Figure 362. Rainbow glacier, 1947.

Figure 363. Rainbow glacier, 1964.

Figure 364. Rainbow glacier, 1987. (Photo by Austin Post)

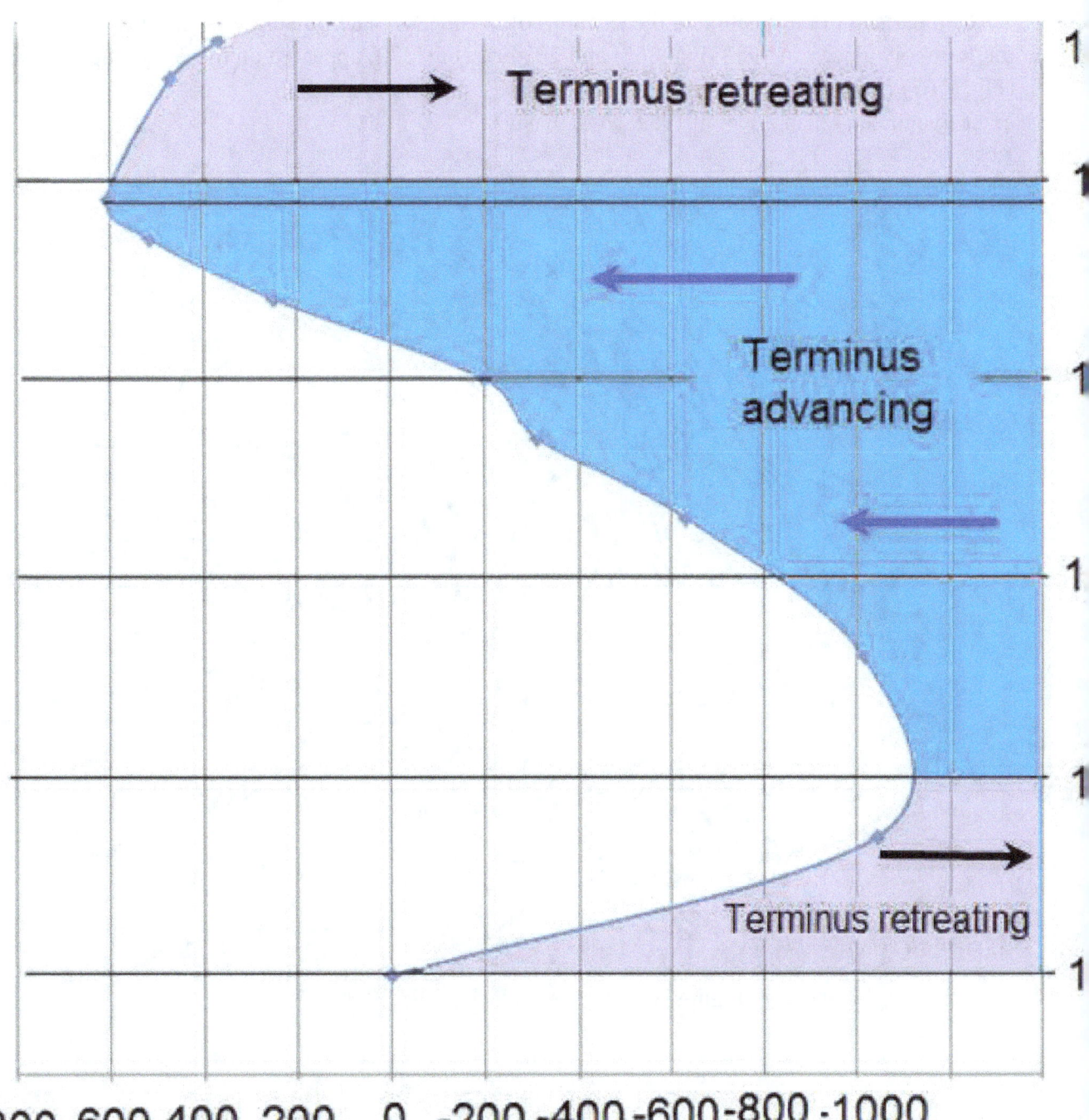

Figure 365. Advance and retreat of the Rainbow glacier.

Comparison of the position of the terminus of the Rainbow glacier in 1952 (Fig. 366) and 2014 (Fig. 367) shows that the glacier was more extensive in 2014 than it was in 1952. Note that a small isolated glacier just north of Lava Divide (Fig. 370) appeared on the 2014 map but was not there in 1952.

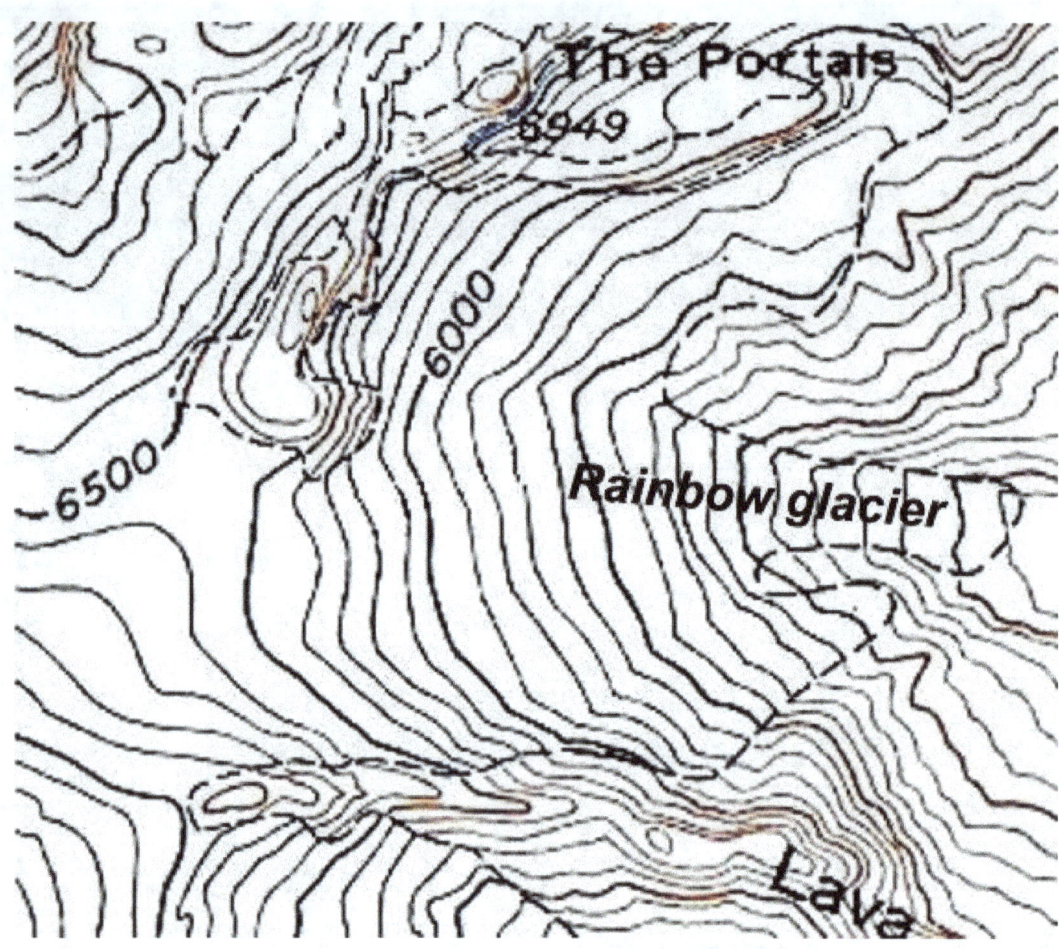

Figure 366. Rainbow glacier, 1952. (USGS map)

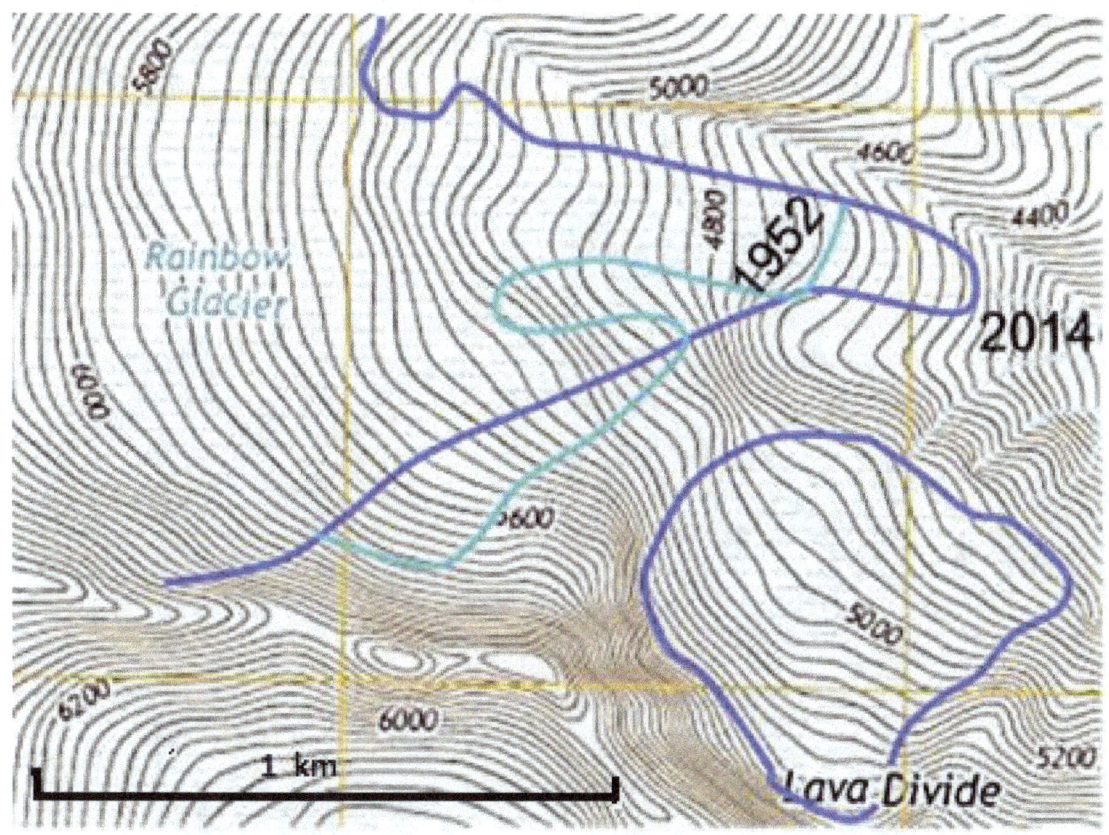

Figure 367. Rainbow glacier, 2014 (blue line). Comparison with 1952 map shows that the glacier was more extensive in 2014 than in 1952. Note the isolated glacier just north of Lava Divide in the 2014. This was not present in 1952. (USGS map)

Mazama glacier

The Mazama glacier originates high on the north flank of Mt. Baker's summit cone and flows into the upper valley of Bar Creek (Fig. 368). Moraines of the glacier have not been studied or directly dated, but a lateral moraine high above the valley floor (Figs. 369, 370) appears to correlate with the Little Ice Age glacier advance elsewhere on the mountain.

Figure 368. Topographic map of the Mazama glacier.

Figure 368. Mazama glacier, 1987. (Photo by Austin Post)

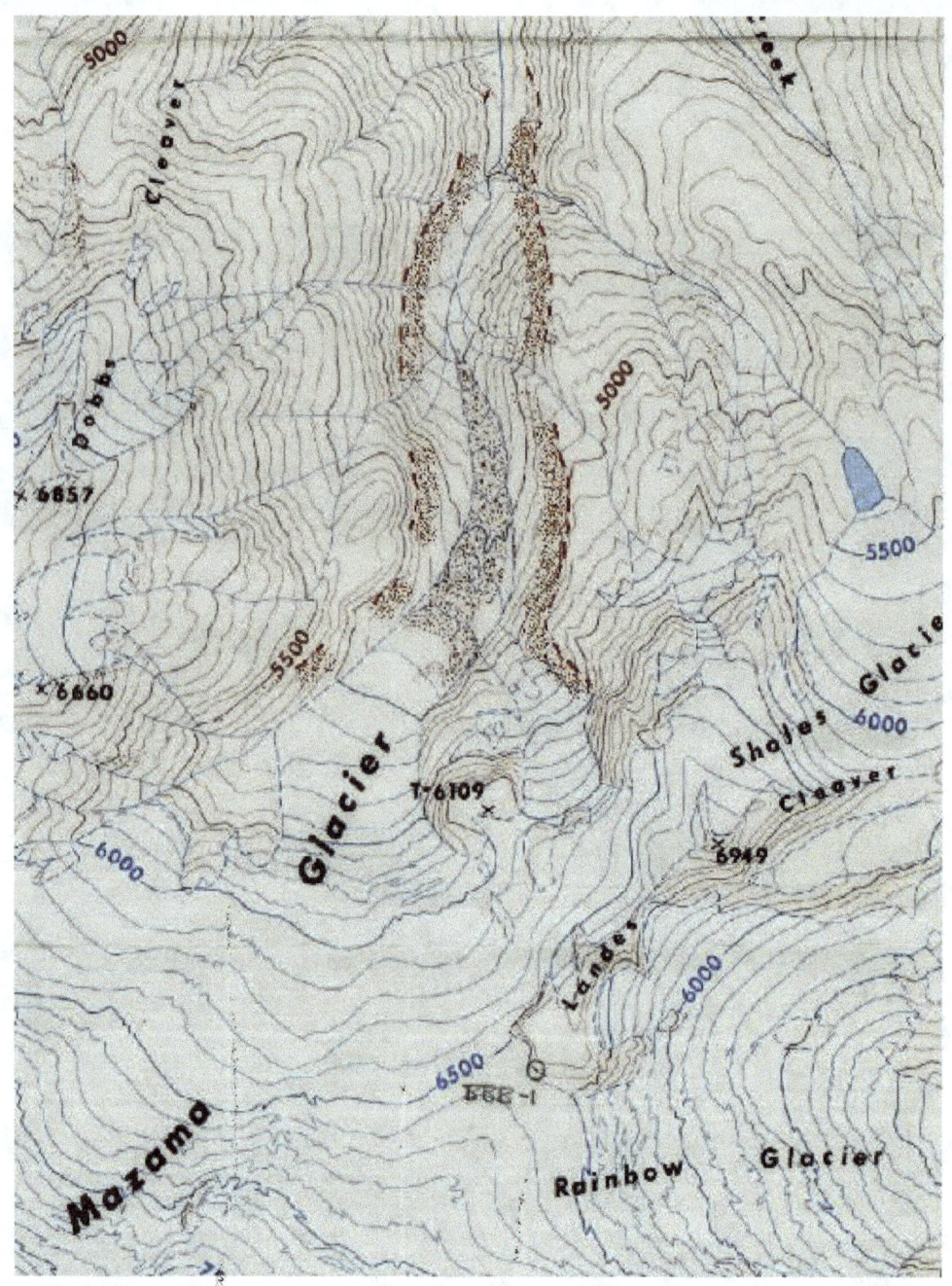

Figure 369. Little Ice Age moraines of the Mazama glacier (brown shading). (Modified from USGS map)

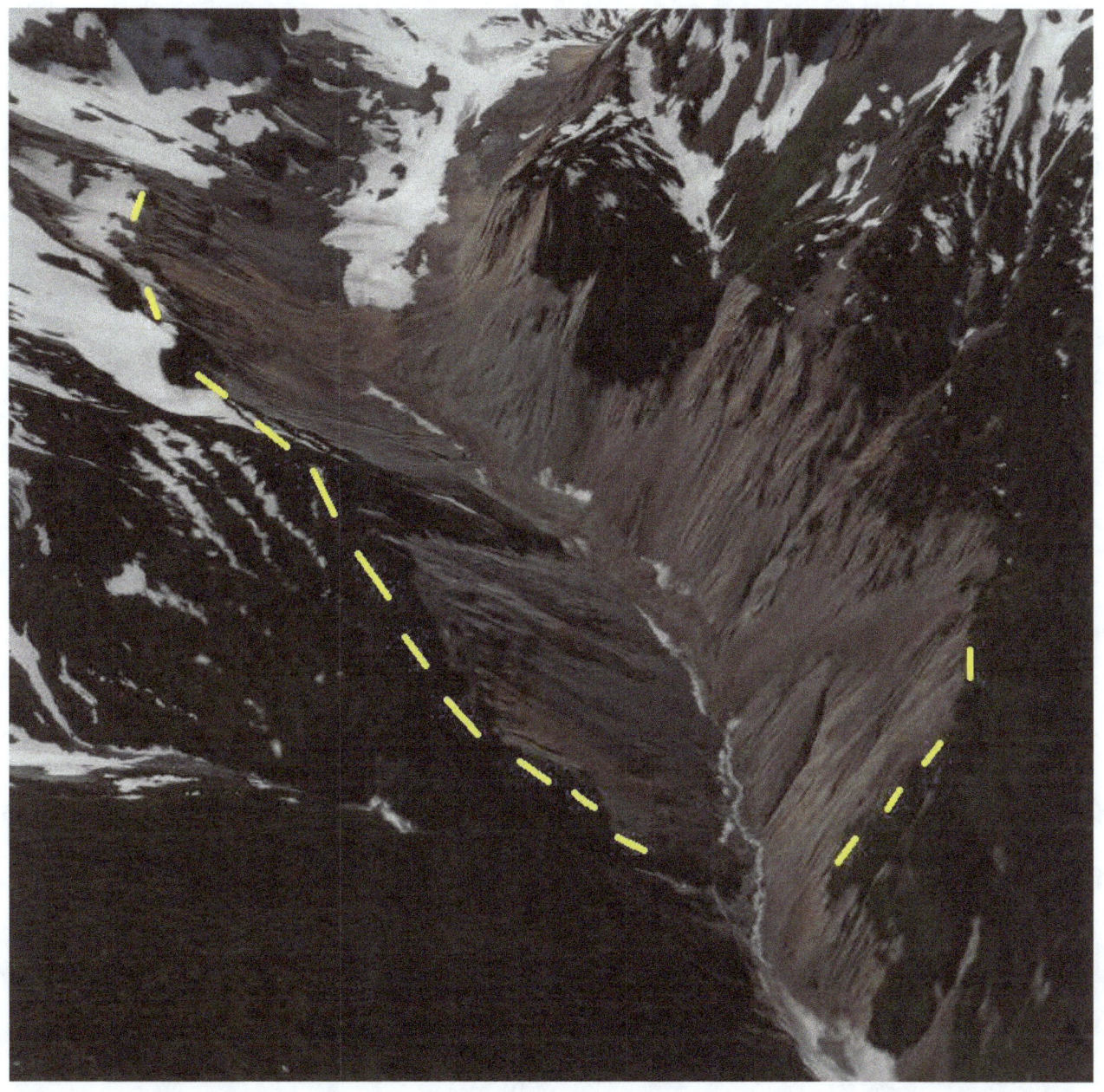

Figure 370. Lateral moraines high above the valley floor appear to be from the Little Ice Age but are not directly dated.

The terminus of the Mazama glacier fluctuated during the past century along with the other Mt. Baker glaciers. From its maximum extent in 1909 (Fig. 371) and 1915 (Fig. 372), the glacier retreated strongly upvalley during the 1915 to 1945 warm period, then advanced 1500 feet from 1950 to 1979 during the 1945 to 1978 cool period. During the 1978 to 2000 warm period, the terminus receded somewhat less than 1500 feet to nearly its position in 1950.

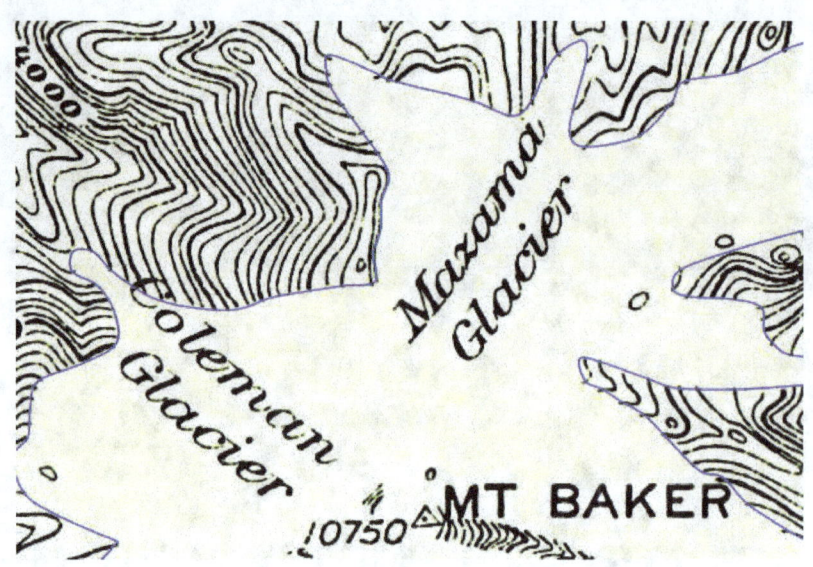

Figure 371. Mazama glacier, 1909. (USGS map)

Figure 372. Mazama glacier, 1915. (USGS map)

Figure 373. Mazama glacier, 1940.

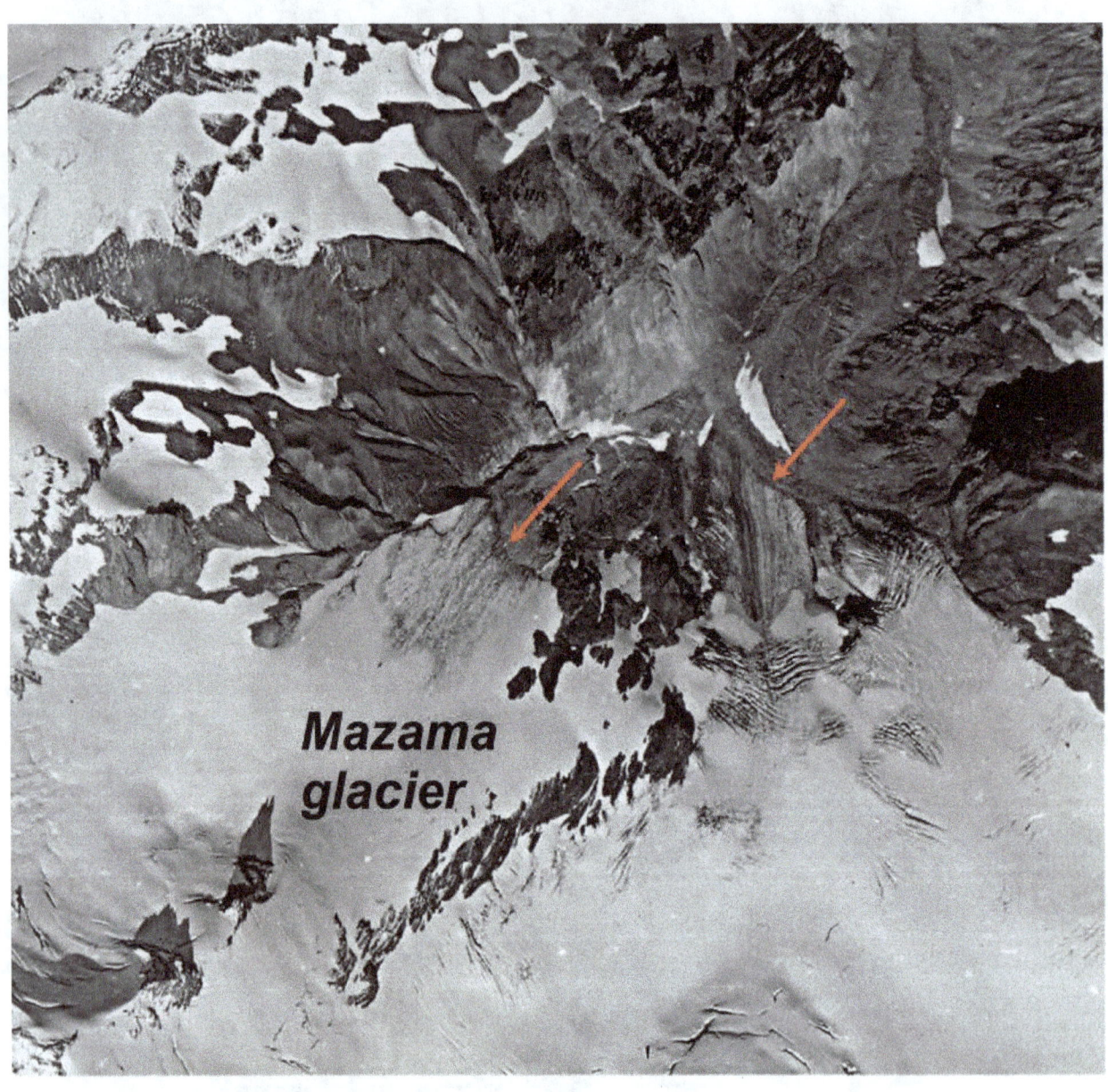

Figure 374. Mazama glacier 1950. Terminus shown by red arrows.
The lower part of the right hand ice lobe has been buried by a large landslide at the red arrow.

Figure 375. Mazama glacier, 1987. (Photo by Ausin Post)

Figure 376. Mazama glacier, 1987.(Photo by Austin Post)

Figure 377. Mazama glacier, 2009.

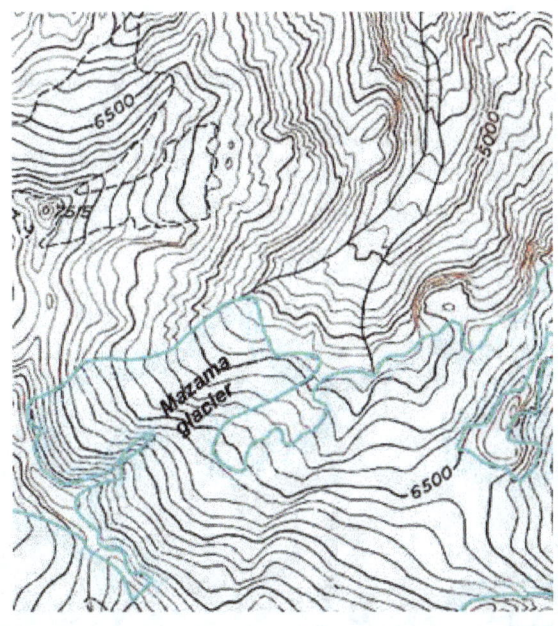

Figure 378. The Mazama glacier in 1952. (USGS map)

Figure 379. Mazama glacier in 2014 (blue). Comparison of 1952 (green) terminus position with 2014 shows that the terminus in 2014 was 1¼ mile farther downvalley than in 1952.

Hadley glacier

The Hadley glacier is a cirque glacier just north of Mt. Baker between the Mazama and Roosevelt glaciers (Fig. 380,381). It is an independent cirque glacier and not connected to the Mt. Baker glacier system, but because it is adjacent to the Mazama and Roosevelt glacies, it provides a useful comparison of cirque and valley glacier response to changes in climate.

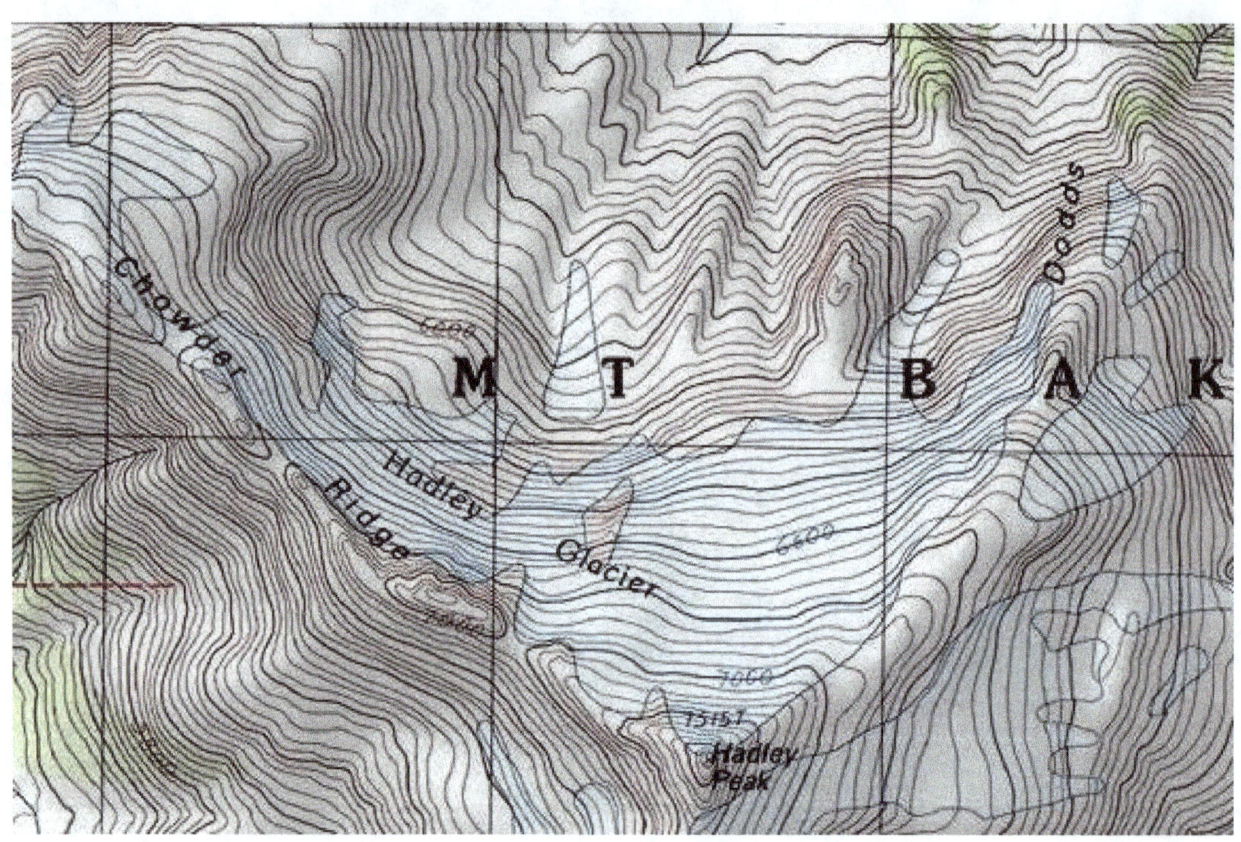

Figure 380. Hadley glacier.

Figure 381. Hadley glacier.

Figure 382 shows the extent of the Hadley glacier during the Little Ice Age (500 yrs ago). The glacier terminus lay well downslope from the modern terminus (brown line).

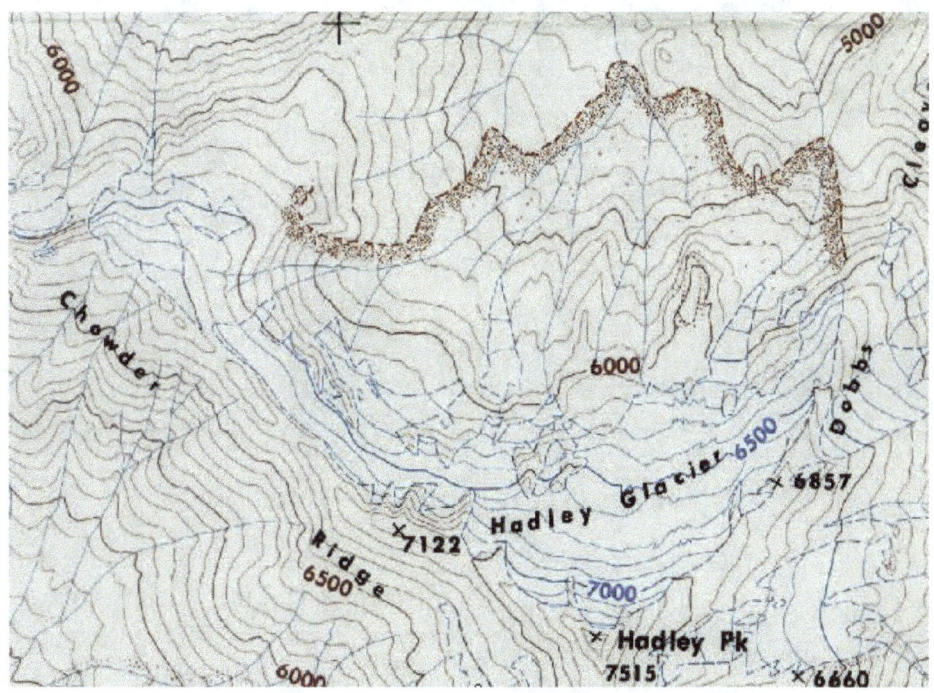

Figure 382. Hadley glacier Little Ice Age terminus.

The Hadley glacier is not shown on the 1909 and 1915 USGS maps so the position of the terminus then is not known. The earliest known photo of the Hadley glacier was taken in 1940 (Fig. 383). By then the glacier must have been in retreat for many years.

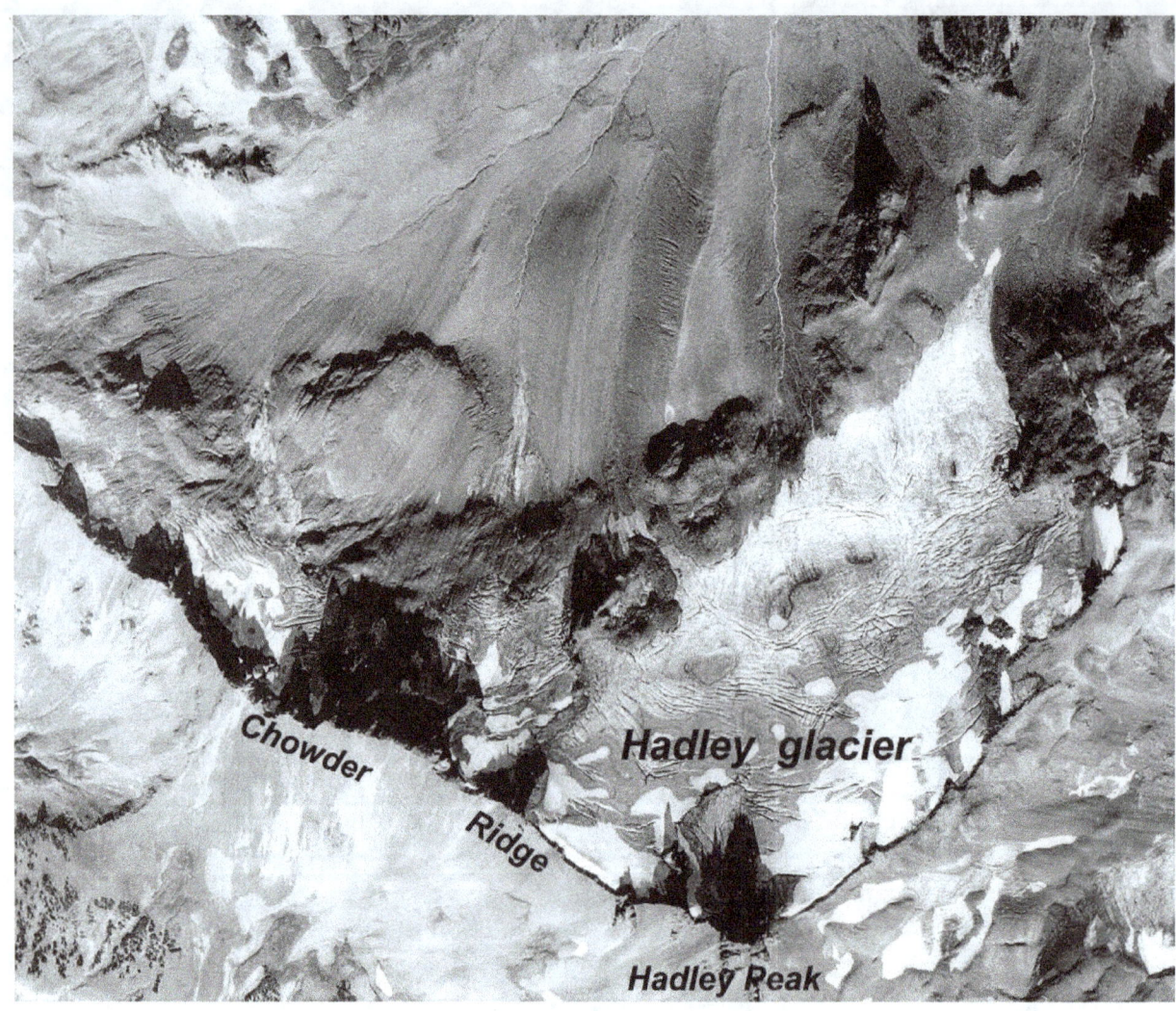

Figure 383. Hadley glacier, 1940.

Figure 384 shows the position of the terminus of the Hadley glacier in 1952 and Figure 385 shows it in 2014. Comparison of the terminus positions on the maps shows very little difference, i.e., the glacier is about the same size now as it was in 1952, unlike all of the glaciers on the main Mt. Baker cone which are all more extensive now than they were in 1952.

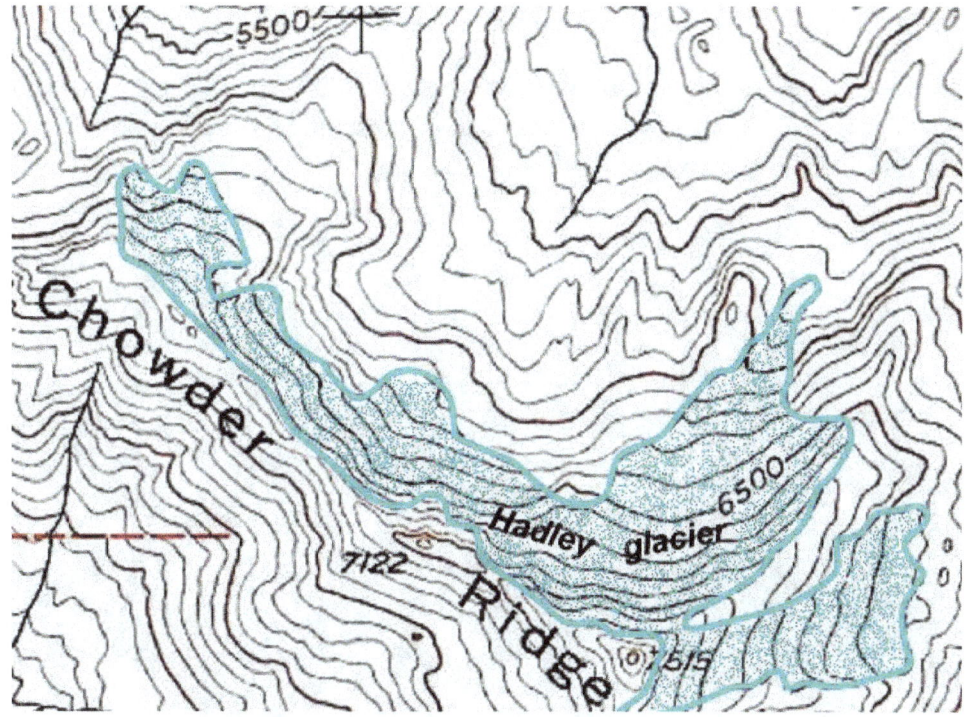

Figure 384. Hadley glacier, 1952. (USGS map)

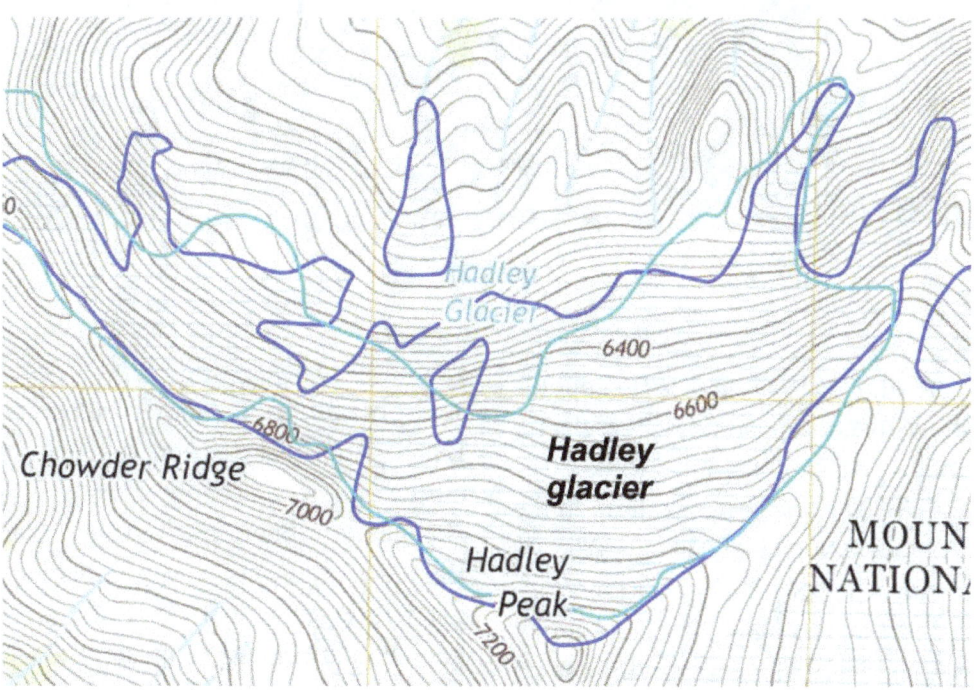

Figure 385. Hadley glacier, 2014 (blue line). Comparison of positions of the 1952 terminus (green line) and 2014 show little difference.

Sholes glacier

The Sholes glacier (Fig. 386) is an independent cirque glacier just adjacent to the Mazama glacier. It is not connected to the Mt. Baker glacier system and is too small to show on the 1909 or 1915 USGS topographic maps so little is known about its pre-1940 history.

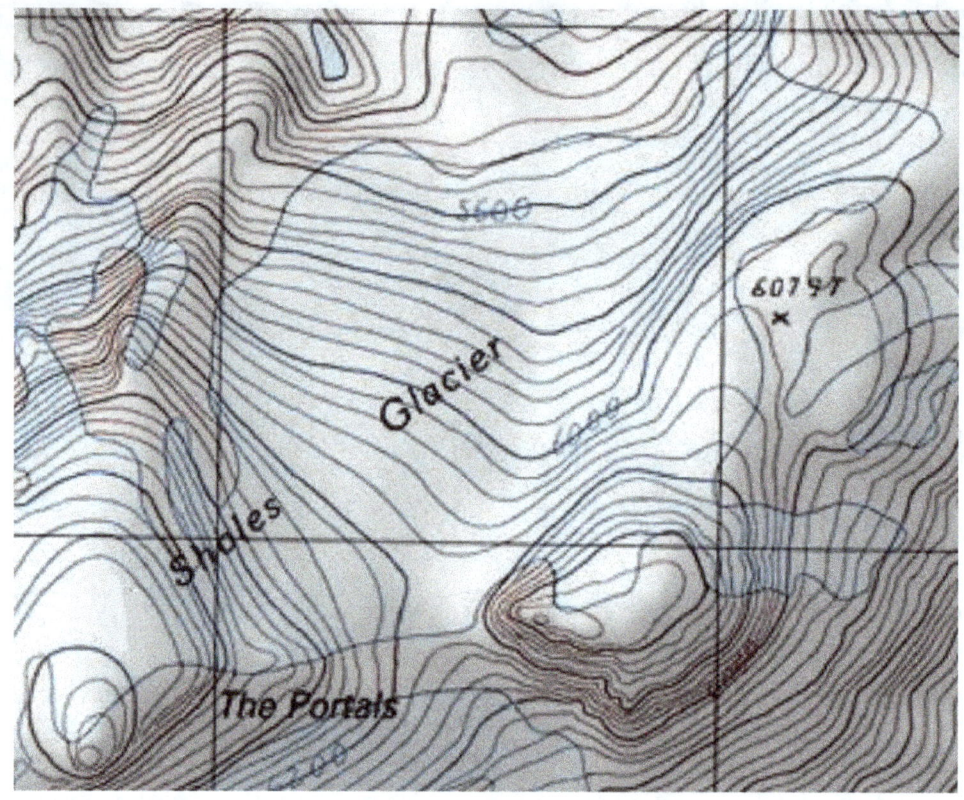

Figure 386. Sholes glacier.

Comparison of the 1947 and 2011 photos (Figs. 387) shows that the position of the glacier terminus in 1947 is virtually identical to its 2011 position (Fig. 388). The terminus position on the 1952 USGS topographic map (Fig. 389) is essentially identical to the 2014 position (Fig. 390). This is the same situation as the Hadley cirque glacier—neither of the two shows the same response to changing climate conditions that strongly affected the larger valley glaciers on Mt. Baker. Thus, it seems apparent that small Cascade cirque glaciers are not nearly as sensitive to changes in climate as the valley glaciers on Mt. Baker, suggesting that fluctuations in cirque glaciers elsewhere in the Cascade may not be good climate change indicators.

Figure 387. Sholes glacier, 1947.

Figure 388. Sholes glacier, 2011.

These photos prove that the Sholes glacier today is almost identical to as in 1947. Comparison of the glacier terminus on USGS topographic maps of 1952 and 2014 (below) shows that the Sholes glacier is slightly larger than it was 1952.

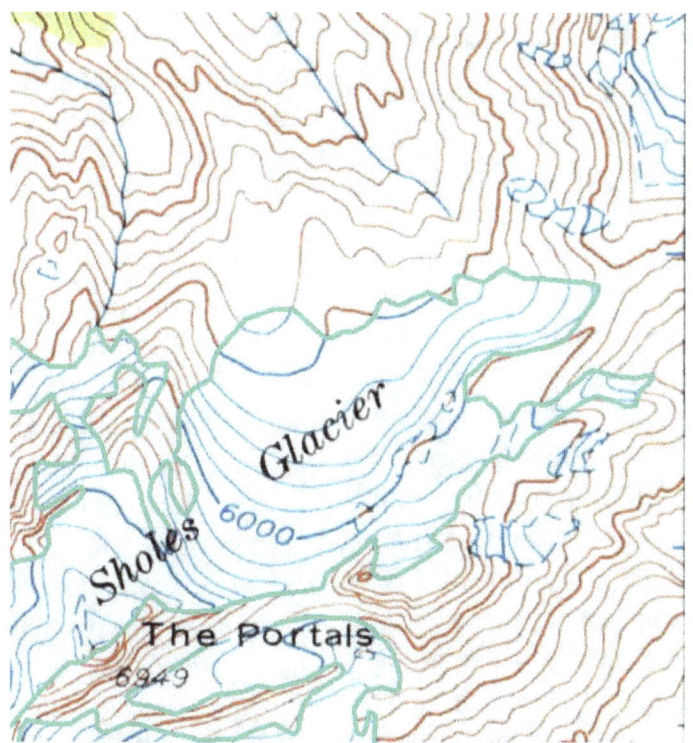

Figure 389. Extent of the Sholes glacier in 1952.

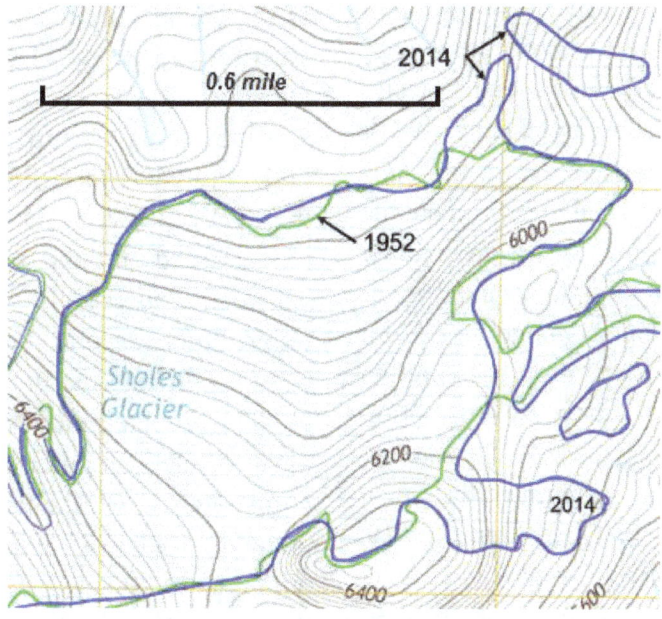

Figure 390. Extent of the Sholes glacier in 2014. The blue line (USGS map) is the margin of the glacier shown on the USGS 2014 map. The green line is the terminus position shown on the 1952 map. These maps prove that the Sholes glacier today is slightly more extensive now than it was in 1952.

Implications of glacial fluctuations on Mt. Baker

Glaciers on Mt. Baker have advanced and retreated as sea surface temperatures in the eastern Pacific Ocean and global temperature warmed and cooled. The Pacific Ocean has switched back and forth between a warm mode and a cool mode for the past 500 years. In its cool mode, like the periods from 1880 to 1915 and 1945 to 1977, ocean temperatures in the eastern Pacific were cool and Mt. Baker glaciers advanced. In its warm mode, like the periods from 1915 to 1945 and 1977 to ~2000, ocean temperatures in the eastern Pacific warmed and Mt. Baker glaciers retreated. The regular mode switches between warm and cool have been going on for more than 500 years and are known as the Pacific Decadal Oscillation (PDO). Each mode typically last 25 to 30 years so Mt. Baker glaciers have advanced and retreated with each mode change. When the eastern Pacific was cool, Mt. Baker glaciers advanced and when it was warm, Mt. Baker glaciers retreated. The glacier fluctuations are clearly driven by changes in the PDO.

The PDO record extends to about 1900, but the glacial record goes back many years and can be used as an indicator of climate changes over many centuries. Extending this ongoing pattern into the future provides an opportunity to predict future climate changes. A 22-year warm period (1978 to ~2000) ended when the PDO flipped into its cool mode in 1999 and the climate has since cooled (Fig. 181). If the PDO behaves as it has through the past four previous global climate changes in the past century, we can look ahead to several decades of global cooling and glacier advance.

Notwithstanding all of the map and air photo evidence above, the front page headline of the September 8, 2015 Seattle Times stated:

'Disastrous': Low snow, heat eat away at Northwest glaciers

"Glaciers across the North Cascades could lose 5 to 10 percent of their volume this year, accelerating decades of steady decline. One scientist estimates the region's glaciers are smaller than they have been in at least 4,000 years." "The best word for it is disastrous," said Maurice Pelto. This story was widely published in many other newspapers and on the internet.

However, the photos and maps of each of Mt. Baker glaciers clearly show that this story was completely untrue. The photos and maps above prove that <u>all</u> Mt. Baker glaciers are more extensive today than they were in 1952, and some of the glaciers are actually advancing. Thus, the assertions made by Maurice Pelto were entirely false.

Even more absurd are the computer model projections by Bob Mitchell and his students in the WWU Geology Dept. He uses a computer model based on the *assumption* that climate will warm continuously 10° F well before the end of this century, despite the facts that (1) there has been no global warming in almost 20 years, (2) all 92 computer models using this assumption have failed miserably to match actual temperatures over the past few decades, (3) there is no real glacial data in the model, and (4) temperatures in the Cascades have cooled over the past decade. Mitchell's model ridiculously projects that 85% of all glaciers on Mt. Baker will be gone by 2100.

REFERENCES

Beget, J.E., 1981, Early Holocene glacier advance in the North Cascade Range, Washington: Geology, vol. 9, p. 409-413.

Burrows, R.A., 2001, Glacial chronology and paleoclimatic significance of cirque moraines near Mts. Baker and Shuksan, North Cascade Range, Washington: M.S.thesis, Western Washington University, 92 p.

Burrows, R., Clark, D.H., Easterbrook, D.J., Kovanen, D., and Slaymaker, O., 2007, Evidence for cirque glacier chronologies and rapid alpine deglaciation in the North Cascades during the Holocene/Pleistocene transition: Geological Society of America, Abstracts with Programs, vol. 39, p. 12.

Burrows, R.A., Kovanen, D.J., Easterbrook, D.J., and Clark, D.H., 2000, Timing of extent of cirque glaciations near Mts. Baker and Shuksan, North Cascades Range, Washington: Geological Society of America, Abstracts with Programs, vol. 32, p. 7.

Crandell, D.R., 1975, Increased hydrothermal activity at Mount Baker: U.S.Geological Survey Professional Paper 975.

Easterbrook, D.J., 1975, Mount Baker eruptions: Geology, vol. 3, p. 679-682.

Easterbrook, D.J., 1976, Pleistocene and Recent volcanic activity of Mt. Baker, Wash.: Abstracts with Programs, Geological Society of America, vol. 8, p.849.

Easterbrook, D.J., 1976, Mt. Baker eruptions: Eos, Transactions, American Geophysical Union, vol. 57, p.87.

Easterbrook, D.J., 1979, The last glaciation of Northwest Washington: Pacific Coast Paleogeography Symposium, Society of Economic Paleontologists and Mineralogists, p.177-189.

Easterbrook, D.J., 1980, Activity of Mt. Baker 1975-1979, Eos, Transactions, American Geophysical Union, vol. 61, p.69.

Easterbrook, D.J., 1986, Stratigraphy and chronology of Quaternary deposits of the.Puget Lowland and Olympic Mountains of Washington and the Cascade Mountains of Washington and Oregon: *in*: Sibrava, V., Bowen, D.Q., Richmond, G.M. eds., Quaternary Glaciations in the Northern Hemisphere, Quaternary Science Reviews, vol. 5, p. 135-159.

Easterbrook, D.J., 1992, Late Quaternary fluctuations of glaciers on Mt. Baker, Wash.: Abstracts with Program, Geological Society of America, vol. 24, p. 21.

Easterbrook, D.J., 1992, Advance and retreat of Cordilleran ice sheets in Washington, U.S.A.: Geographie Physique et Quaternaire, vol. 46, p. 51-68.

Easterbrook, D.J., 1999, Surface processes and landforms: Prentice-Hall, 546 p.

Easterbrook, D.J., 2002, Pleistocene glaciation of western Washington: Abstracts with Program, Geological Society of America, vol. 34, p.108.

Easterbrook, D.J., ed., 2003, Quaternary Geology of the United States: International Quatenary Association, 2003 Field Guide Volume, Desert Research Institute, Reno, NV, 438 p.

Easterbrook, D.J., 2003, Cordilleran Ice Sheet glaciation of the Puget Lowland and Columbia Plateau and alpine glaciation of the North Cascade Range, Wash.: in Easterbrook, D.J., ed., Quaternary Geology of the United States, International Quaternary Association, 2003 Field Guide Volume, Desert Research Institute, Reno, NV, p. 265-286

Easterbrook, D.J., 2003, Cordilleran Ice Sheet glaciation of the Puget Lowland and Columbia Plateau and alpine glaciation of the North Cascade Range, Washington: Geological Society of America Field Guide 4, p. 137–157.

Easterbrook, D.J., 2007, Historic Mt. Baker glacier fluctuations—geologic evidence of the cause of global warming: Abstracts with Program, Geological Society of America, vol. 39, p.13.

Easterbrook, D.J., 2007, Younger Dryas to Little Ice Age glacier fluctuations in the Fraser Lowland and on Mt. Baker, Washington: Geological Society of America, Abstracts with Programs, vol. 39, p. 11.

Easterbrook, D.J., 2010, A walk through geologic time from Mt. Baker to Bellingham Bay: Chuckanut Editions, Village Books, 329 p.

Easterbrook, D.J., 2015, Late Quaternary glaciation in the Puget Lowland, North Cascade Range, and Columbia Plateau, Washington: University of Washington Press, p. 257-286.

Easterbrook, D.J., and Burke, R.M., 1972, Neoglaciation of the northern Cascades,Washington: Geological Society of America, Abstracts with Programs, vol. 4, p. 152.

Easterbrook, D. J. and Burke, R. M., 1972, Neoglaciation on the flanks of Mount Baker, Washington: Northwest Science Assoc. Abstracts, Pullman, Washington.

Easterbrook, D.J., and Burke, R.M., 1972, Glaciation of the northern Cascades, Wash.: Geological Society of America, Abstracts with Program, v. 4, p. 152.

Easterbrook, D.J., and Donnell, C.B., 2007, Glacial and volcanic history of the Nooksack Middle Fork, Washington: Geological Society of America, Abstracts with Programs, vol. 39, p. 12.

Easterbrook, D.J. and Kovanen, D.J., 1996, New evidenc for late-glacial, post- Cordilleran ice-sheet readvance of alpine glaciers in the North Cascades, Washington: Abstracts with Programs, Geological Society of America, vol. 28, p.83.

Easterbrook, D.J., and Kovanen, D.J., 1996, Evidence for 45-km-long, post- Cordilleran-Ice-Sheet, alpine glaciers in the Nooksack North Fork, North Cascades, WA, between 11,500 and 10,000 ^{14}C yrs BP: Geological Society of America, Abstracts with Programs, vol. 28, p. 434.

Easterbrook, D.J. and Kovanen, D. J., 1996, Far-reaching mid-Holocene lahar from Mt. Baker in the Nooksack Valley of the North Cascades, Washington: Abstracts with Programs, Geological Society of America, vol. 28, p.64

Easterbrook, D.J., and Kovanen, D.J., 1999, Early Holocene glaciation of the North

Cascades near Mt. Baker, Washington: Geological Society of America, Abstracts with Programs, vol. 31, p. 367.

Easterbrook, D.J. and Kovanen, D.J., 2000, Cyclical oscillations of Mt. Baker glaciers in response to climatic changes and their correlation with periodic oceanographic changes in the Northeast Pacific Ocean: Abstracts with Programs, Geological Society of America, vol. 32, p.17.

Easterbrook, D.J., Kovanen, D.J., and Slaymaker, O., 2007, New developments in Late Pleistocene and Holocene glaciation and volcanism in the Fraser Lowland and North Cascades, Washington: Geological Society of America Field Guide 9, p. 31-56.

Easterbrook, D.J. and Rahm, D.A., 1970, Landforms of Washington: Union Printing Co., 156 p.

Frank, D., Post, A., and Friedman, J.D., 1975, Recurrent geothermally induced debris avalanches on Boulder Glacier, Mount Baker, Washington: U.S. Geological Survey Journal of Research, 3, p. 77-87.

Fuller, S.R., 1980. Neoglaciation of Avalanche and the Middle Fork Nooksack River Valley, Mt. Baker, Washington: M.S. thesis, Western Washington University, 68 p.

Fuller, S.R., Easterbrook, D.J., and Burke, R.M., 1983, Holocene glacial activity in five valleys on the flanks of Mt. Baker, Washington: Geological Society of America Abstracts with Programs, vol. 15, p. 430-431.

Green, N.L., 1988, Basalt-basaltic andesite mixing at Mount Baker volcano, Washington: Estimation of mixing conditions: Journal of Volcanology and Geothermal Research, vol. 34, p. 251-265.

Heikkinen, O., 1984, Dendrochronological evidence of variations of Coleman Glacier, Mount Baker, Washington, U.S.A: Arctic and Alpine Research, vol. 16, p. 53-64.

Haldreth, W., Fierstein, J., and Lanphere, M., 2003. Eruptive history and geochronology of the Mount Baker volcanic field, Washington: Geological Society of America Bulletin, vol. 115, p. 729-764.

Hyde, J.H., and Crandell, D.R., 1978, Postglacial volcanic deposits at Mount Baker, Washington, and potential hazards from future eruptions: U.S. Geological Survey Professional Paper 1022-C, 17 p.

Kovanen, D.J., and Begét, J.E., 2005, Comments on "Early Holocene glacier advance, southern Coast Mountains, British Columbia." Quaternary Science Reviews, vol. 24, p. 1521-1526.

Kovanen, D.J., and Easterbrook, D.J., 1996, Extensive readvance of Late Pleistocene (YD?) alpine glaciers in the Nooksack River Valley, 10,000 to 12,000 years ago, following retreat of the Cordilleran Ice Sheet, North Cascades, Washington: *in*:Friends of the Pleistocene Field Trip Guidebook, 74 p.

Kovanen, D.J. and Easterbrook, D.J., 1998, 10,600-yr-old age of post-Cordilleran ice-sheet alpine glaciation in the Nooksack Valley, North Cascades, WA: Abstracts with Programs, Geological Society of America, vol. 30, p.165.

Kovanen, D.J., and Easterbrook, D.J., 1999, Holocene tephras and lahars from Mt. Baker, Washington: Geological Society of America, Abstracts with Programs, vol. 31, p. 71.

Kovanen, D.J., and Easterbrook, D.J., 2001, Late Pleistocene, post-Vashon, alpine glaciation of the Nooksack drainage, North Cascades, Washington: Geological Society of America Bulletin, vol. 113, p. 274-288.

Kovanen, D.J., and Easterbrook, D.J., 2002. Timing and extent of Allerød and Younger Dryas age (ca.12,500-10,000 14C yr B.P.) oscillations of the Cordilleran ice sheet in the Fraser Lowland, western North America: Quaternary Research, vol. 57, p. 208-224.

Kovanen, D.J., and Slaymaker, O., 2005. Fluctuations of Deming Glacier and theoretical equilibrium-line altitudes during the late Pleistocene and early Holocene on Mount Baker, Washington, USA: Boreas, vol. 34, p. 157-175.

Kovanen, D.J., Easterbrook, D.J., and Thomas, P.A., 2001, Holocene eruptive history of Mount Baker, Washington: Canadian Journal of Earth Sciences, vol. 38, p. 1355-1366.

Porter, S.C., and Swanson, T.W., 2008, ^{36}Cl dating of the classic Pleistocene glacial record in the northeastern Cascades Range, Washington: American Journal of Science, vol. 308, p. 130-166.

Post, A., Richardson, D., Tangborn,W., and Rosselot, F.L., 1971, Inventory of glaciers in the North Cascades: U.S. Geological Survey Professional Paper 705-A.

Riedel, J.L., 2007, Late Pleistocene glacial and environmental history of Skagit Valley, Washington and British Columbia: Ph.D. thesis, Simon Fraser University, Burnaby, British Columbia, 187 p.

Stuiver, M., Reimer, P.J., and Reimer, R.W., 2005, CALIB 5.0.2 radiocarbon calibration program. http://calib.qub.ac.uk/calib/.

Thomas, P.A., 1997, Late Quaternary Glaciation and Volcanism on the South Flank of Mt. Baker, Washington: M.S. thesis, Western Washington University, 98 p.

Thomas, P.A and Easterbrook, D.J., 1997, Late Quaternary glacial advances on Mt Baker, Washington: Abstracts with Programs, Geological Society of America, vol. 29, p.69.

Thomas, P.A., Easterbrook, D.J., and Clark, P.U., 2000, Early Holocene glaciation on Mt. Baker, Washington state, USA: Quaternary Science Reviews, vol. 19, p. 1043-1046.

Westgate, J. A., Easterbrook, D. J., Naeser, N. A., and Carson, R. J., 1987, The Lake Tapps tephra: an early Pleistocene stratigraphic marker in the Puget Lowland, Washington: Quaternary Research, vol. 28, p. 340-355.

GLOSSARY

Ablation area: That part of a glacier or snowfield where ablation exceeds accumulation.

Ablation: The combined processes by which a glacier wastes.

Abrasion: The wearing away by friction.

Accumulation area: The area of a glacier in which annual accumulation exceeds ablation.

Aggradation: The process of building up a surface by deposition.

Alluvial fan: Low, cone-shaped deposit formed by a stream issuing from mountains into a lowland.

Alluvium: Sand, gravel, and silt deposited by rivers and streams in a valley bottom.

Alpine glacier: Glaciers occupying mountainous terrain.

Altitude: The vertical distance between a point and mean sea level.

Amphibole: A family of silicate minerals forming prismatic or needlelike crystals in many igneous and metamorphic rocks. Actinolite is a light green form of amphibole.

Andesite: Fine-grained, generally gray to dark-colored, volcanic rock. Commonly has visible crystals of plagioclase feldspar. Occurs in lava flows and dikes. The most common rock in volcanic arcs.

Anticline: Arched geologic structure in which beds dip in opposite directions from the central axis.

Arete: A sharp-crested mountain ridge between two cirque headwalls.

Argillite: Fine-grained sedimentary rocks made mostly of silt and clay. Includes shale, mudstone, siltstone, and claystone. Commonly black.

Basal till: Poorly sorted mixture of sand, silt, clay, pebbles, cobbles, and boulders deposited from, the base of a glacier.

Basalt: Fine-grained, generally black, volcanic rock relatively rich in iron, magnesium, and calcium. Occurs in lava flows and dikes.

Batholith: Very large mass of slowly cooled, intrusive molten rock, such as granite, at least 50 square miles in area.

Bedding: Sedimentary layers in a rock. Beds are distinguished from each other by grain size and composition.

Blueschist: Metamorphic rock rich in blue amphibole formed by high pressure and low metamorphic temperature. Makes up much of Mt. Shuksan.

Breccia: Rock made up of angular fragments of other rocks. **Volcanic breccia** is made of volcanic rock fragments generally blown out of a volcano or eroded from it.

Calcite. Mineral made of calcium carbonate ($CaCO_3$). Generally white, easily scratched with knife. Most seashells are calcite. Primary component of limestone.

Caldera: A large circular volcanic depression, the diameter of which is many times greater than that of normal volcanic cones. Caused by collapse of a volcanic cone upon the withdrawal of magma from below.

Carbon-14: A radioactive isotope of carbon with atomic weight 14, produced by collisions between neutrons and atmospheric nitrogen in the upper atmosphere. Used as a natural geologic clock to determine the age of organic material, such as wood or shells.

Chert: Sedimentary rock made of fine-grained quartz. Usually made of millions of globular siliceous skeletons of tiny marine plankton called radiolarians.

Chlorite: Family of green, platy, silicate minerals common in low grade metamorphic rocks

Cinder cone: A volcanic cone formed by the accumulation of volcanic cinders or ash around a vent.

Cirque: A deep, steep-walled recess in a mountain caused by glacial erosion.

Clay: Particles less than 1/16 millimeter in diameter. Also a family of platy silicate minerals generally too small to be seen even with a microscope. A common product of rock weathering, especially of rocks containing much feldspar. The term *clay is* also used to refer to very, very fine sedimentary grains whether or not they really are made of clay minerals.

Climate: The sum total of the meteorological elements that characterize the average and extreme condition of the atmosphere over a long period of time at any one place or region of the earth's surface.

Composite cone: A volcanic cone, usually of large dimension, built of alternating layers of lava and fragmental material.

Conglomerate: Sedimentary rock made of rounded pebbles, cobbles, and boulders greater than at least 2 mm (about 1/13th of an inch) in diameter.

Continental glacier: A large ice sheet covering a large part of a continent.

Contour interval: The difference in elevation between two adjacent contour lines.

Contour: An imaginary line on the surface of the ground, every point of which is at the same altitude.

Cordilleran Ice Sheet: Ice cap that grew in western North America during the Pleistocene Epoch, beginning in Canada and covering much of British Columbia, Alaska, and northernmost United States.

Crevasse: A fracture in glacial ice formed under tensional stress in the ice.

Debris flow: A moving mass of water-lubricated debris.

Deglaciation: The uncovering of an area from beneath glacier ice as a result of shrinkage of a glacier.

Delta: An alluvial deposit, often triangular-shaped, formed where a stream enters the ocean or a lake and drops its load of sand, silt, or gravel.

Differential erosion: The more rapid erosion of portions of the earth's surface as a result of differences in the erodibility of the rock or in the intensity of surface processes.

Differential weathering: When rocks are not uniform in character but are softer or more soluble in some places than in others, an uneven surface may be developed.

Dike. Tabular body of igneous rock formed where molten rock fills a crack in preexisting rock.

Diorite. Intrusive igneous rock made of plagioclase feldspar and amphibole and/or pyroxene. Similar to granite except it has little or no quartz.

Dip: The angle at which a bed or other planar feature is inclined from the horizontal.

Discharge: Rate of flow at a given instant in terms of volume per unit of time through a given cross-sectional area.

Divide: The line of separation between drainage systems; the summit of a ridge between streams.

Drainage basin: The area drained by a river system.

End moraine: A ridge-like accumulation of glacial sediment deposited at the terminus of a glacier.

Erratic: A rock transported by a glacier or by floating ice, different from the bedrock on which it lies

Escarpment: A cliff or relatively steep slope separating gently sloping tracts.

Facet: A flat surface produced by abrasion on a rock.

Fault scarp: A bluff formed by a fault offsetting the land surface.

Fault: A fracture along which there has been displacement of the two sides relative to one another. Abrupt movements on faults cause earthquakes. Where the crack is roughly vertical the rocks may move up or down or sideways or in some combination. If the fault is inclined at a low angle to the Earth's surface and rock on one side of the fault moves up and over rock on the other side, it is a **thrust fault.**

Feldspar: Family of silicate minerals whose crystals are, generally white (plagioclase) or pink (potassium feldspar). The most abundant mineral in the Earth's crust.

Fission tracks: Microscopic tunnels in crystals and glass made by nuclear particles emitted by radioactive elements, usually uranium. The number of fission tracks in glass and zircon crystals increases with time and can be used as a dating method..

Foliation. Parallel arrangement of minerals, especially platy minerals such as micas, in a rock, so as to give it a foliated look, like pages in a book. Foliated rocks tend to break along the foliation and form slabs. Mostly found in metamorphic rocks.

Fluting: Smooth deep furrows worn in the surface of rocks by glacial or stream erosion.

Fluvial: Pertaining to rivers or produced by river action.

Geomorphology: The study of physical and chemical processes that affect the origin and evolution of surface landforms.

Glacial drift: All sediment deposited directly or indirectly from a glacier or by its meltwater.

Glacial striae: Scratches on smoothed surfaces of rocks made by glacial erosion.

Glacial trough: U-shaped valley shaped by glacial erosion.

Glacier: A body of ice, firn, and snow, originating on land and showing evidence of past or present movement.

Gneiss: A light-colored, coarse grained metamorphic rock made by recrystallization of older rocks with chemical compositions similar to granite.

Gradient: Slope expressed as the angle of inclination from the horizontal.

Granite: A coarse-grained igneous rock made of feldspar and quartz that has crystallized from molten rock at great depth below the Earth's surface where crystallization is slow and the minerals are large.

Granitic rocks: A general term for coarse grained igneous rocks composed mostly of feldspar with or without quartz.

Greenstone: A green metamorphic rock made by low temperature, high pressure recrystallization of basalt. Greenstones contain the green minerals chlorite, actinolite, and epidote, which make the rock green. Makes up much of Mt. Shuksan

Groundwater: Subsurface water.

Horn: A high pyramidal peak with steep sides, formed by the intersecting walls of several cirques, as the Matterhorn in Switzerland. The summit pyramid of Mt. Shuksan is a horn

Hornblende: A specific type of the mineral group amphibole. Usually black or very dark green.

Ice sheet: A large glacier of continental proportions forming a continuous cover over a land surface.

Igneous rocks: Rocks formed by crystallization of molten rock.

Interglacial: Pertaining to the time between glaciations.

Intrusion: Injection of molten rock into other preexisting rocks.

Isotope: Elements with slightly different numbers of neutrons in their nucleus than is usual for a particular element. For example, radiocarbon (^{14}C) is a radioactive isotope of carbon with 14 neutrons in its nucleus, rather than the normal 12.

Joint: A fracture in a rock.

Kame terrace: A terrace of glacial sand and gravel deposited between a glacier and the valley sides.

Kame-and-kettle topography: Surface formed by a kame complex interspersed with kettles.

Kettle: A depression in glacial sediments made by the melting of a detached mass of glacier ice that has been either wholly or partly buried by sediment.

Lacustrine: Pertaining to lakes.

Lahar: Volcanic mudflow

Landslide: Any downhill sliding of earth.

Lateral moraine: An elongate ridge of glacial debris deposited along the sides of a glacier.

Lava: Molten rock that has flowed out onto the Earth's surface.

Limestone: A sedimentary rock composed of the mineral calcite (calcium carbonate--$CaCO_3$) commonly formed from the calcium carbonate shells of marine creatures

Little Ice Age: Period of global cold climate from about 1300 AD to the 20th century.

Magma: Molten rock formed below the surface of the Earth. When magma pours out on the Earth's surface it is called **lava.**

Mass wasting: The downslope movement of rock debris under the influence of gravity.

Metamorphic rocks: Rocks formed by recrystallization of older rocks by heat and pressure.

Mica: Group of flat, plate-like silicate minerals, which cleave into smooth, flat flakes. **Biotite** is black. **Muscovite** is light-colored.

Medieval Warm Period: A period of global warm climate from about 900 to 1300 AD

Mineral: A naturally occurring, crystalline compound having definite chemical and physical properties

Moraine: A ridge of glacial sediment along the margin of a glacier composed of rock debris. An **end moraine** forperms at the terminus of a glacier.

Morainal lake: Lake formed behind a morainal dam

Mudflow: Viscous flowage of mud and sediment lubricated with water.

Olivine: A green, glassy mineral formed at high temperature. Common in basalt, especially ocean-floor basalt.

Outwash. Glacial deposit of sand, silt, and gravel formed downstream from a glacier by meltwater streams and rivers.

Outwash plain: A topographic plain made by deposition of sand and gravel by meltwater streams from a glacier.

Peridotite: Rock made of olivine and pyroxene.

Phyllite: Fine-grained, foliated metamorphic rock, generally derived from shale or fine-grained sandstone. Phyllites are usually black or dark gray; the foliation is commonly crinkled or wavy. Differs from less recrystallized slate by its sheen, which is produced by barely visible flakes of muscovite (mica).

Pleistocene: The last Ice Age.

Pluton: Body of igneous of rock that crystallized from molten rock deep in the earth.

Pyroxene: Family of dark green silicate minerals common in basalt and gabbro.

Quartz: Glassy-looking silicon dioxide (SiO_2). One of the most common minerals in the Earth's crust. Found in granite, veins, and sandstone.

Radiocarbon age: The age of organic material determined by the amount of the radioactive carbon ^{14}C in a sample.

Radiocarbon years: Age determined by analysis of carbon-14. Somewhat younger than calendar years.

Recessional moraine: End moraine formed by a stillstand of ice during recession of a glacier.

Reverse fault: A fault in which one side has moved up the fault plane.

Rhyolite: A volcanic rock chemically equivalent to granite but erupted on the land surface. Usually light-colored and fine-grained with tiny, visible crystals of quartz in a dense matrix.

Scarp: A cliff or steep slope.

Schist: Conspicuously foliated metamorphic rock usually derived by recrystallization from shale by heat and pressure.

Sedimentary rocks: Rocks formed deposition of rock particles that are later cemented together to form rock, by precipitation of chemicals in oceans or lakes, or by accumulation of shells or other organic material.

Serpentine: Low-temperature metamorphism of minerals in ultramafic rocks to form green, greasy-looking, silicate minerals that are slippery to the touch.

Shale: Sedimentary rock form by deposition of mud on the floor of oceans or lakes.. Commonly bedded.

Silicate: SiO_4, molecule that is the fundamental building block of **silicate minerals**, consisting of one atom of silica bonded to 4 atoms of oxygten. Silicate minerals make up most rocks at the Earth's surface.

Slump: The downward slipping of a mass of rock or unconsolidated material, usually with backward rotation along a concave-up plane of failure.

Stagnant ice: Glacial ice that has ceased to move.

Striations: Scratches or small grooves.

Subglacial: Beneath a glacier.

Syncline: A trough-like fold in which the beds dip inward from both sides toward the central axis.

Talus: An accumulation of loose rock at the base of a cliff.

Tarn: Alpine lake in the floor of a cirque caused by erosion of the cirque bedrock floor or by a moraine dam.

Terminal moraine: A ridge of glacial deposits marking the farthest advance of a glacier.

Terrace. Flat, gently inclined, or horizontal surface bordered by an escarpment. Can be either depositional or erosional.

Thrust fault: A low-angle, reverse fault that pushes older rocks over younger rocks.

Thrust plate: Slab of rock, generally on the scale of a mountain or greater, bounded by one or more thrust faults.

Till: Poorly sorted, nonstratified rock debris deposited by a glacier.

Ultramafic rock: Rock very rich in pyroxene and olivine. Igneous varieties are peridotite and dunite. May come from the Earth's mantle. A common metamorphic variety is serpentinite.

Vein: Tabular rock filling of fractures in rock with minerals precipitated from hot solutions.

Volcanic arc: Arcuate chain of volcanoes formed above a subducting plate. The arc forms where the descending plate   mostly volcanic rocks from the volcanoes and sedimentary- rocks made up of eroded debris from the volcanoes.

Volcanic ash: Volcanic rock fragments, glass, pumice, and mineral crystals

Volcanic rocks: Rocks formed at the Earth's surface by the solidification of molten rock.

Weathering: Disintegration and decomposition of rocks by surface processes.

Younger Dryas: A glacial period between 11,000 and 12,500 years ago at the end of the last Ice Ice.

Zircon: Mineral composed of zirconium, silicon, and oxygen in igneous rocks.

INDEX

Arbuthnet Lake   30,41,32

Artist Point   13,19,29,30,33,40,41,42,47-49,52,88,89,117,128

Ash   6,8,10,-15,35,39,40,42,58,71,79,80-92,98,100,109,110

Ashflows   11,13,15,19,48,58

Austin Pass   6,31,33,50,133

Bagley Lakes   29,32,40,45,46,128-135

Baker Lake   75-79,262

Baker Pass   21,62

Bar Creek   15,62,302

Barometer Mt.   10

Bastille Ridge   21

Black Buttes   8-10,20-30,46,58,65,139,166,173,176,202

Boulder glacier   260-280

Boulder ridge   60,61

Cathedral Crag ash   21,39,81,85,87-90

Chain Lakes   15,18,19,30,34,41,42,46,88,128,129,133

Chowder Ridge   15,16

Coleman glacier   139-166

Coleman Pinnacle   18,19,29,42,46,47

Colfax Peak   21-28

Columnar jointing   21,33

Cordilleran Ice Sheet   13,40,47,115-118,124,126,129,138,203

Crag View   62

Cougar Divide   10,14-17

Curtis glacier   50,56

Debris avalanches   105-108

Deming glacier   21,62,166,173-199

Dikes   10,13,15,19

Easton glacier   75,200-220

Easton shelf glacier   221-233

Erratics   116,117,121

Glacial grooves   117

Glacial polish   117

Granite   48,49,57,124

Grant Peak (summit cone)   20, 58,59,63-71

Hadley glacier   312-315

Hayes Lake   41,43,46

Heliotrope Ridge   20,142,166

Heather Meadows   29,87,32-40,88,89

Highwood Lake   128,129,131,132

Iceberg Lake   42-44,46,133,135

Kendall moraine   124,125

Kulshan caldera   8-15,18,48

Kulshan ridge   29,47

Lahars   109-113

Lake Ann intrusion   48-50,57

Lake Ann trail   50

Lake Tapps ash   13,14

Lake Whatcom   8,120,126,127

Lasiocarpa Ridge   17,18

Lincoln Peak   19-26,202

Little Ice Age   137,138,141-145,167, 180,183, 185,188,205,206,222-224,228-230,236,237, 249,251,262,265,266,282,293,302,304,305,313

Lookout Mt.   15

Mazama ash   35,40,42,85,87-90,110,130,221,225,228,229,262

Mazama glacier   300-311

Meadow Point   21

Mosquito Lake   120-123

Mt. Herman   40,45,117

Mt. Shuksan   31,35-38,46,48-57,120,124, 128,130

Nooksack Alpine Glacier System   118-127

Panorama Dome   29,48

Park Butte   136,226,234

Park glacier   281-290

Picture Lake   35-37,39,40,89,128-131

Ptarmigan Ridge   17,19,20,29,30,42,46-48,88

Quimper   6

Rainbow glacier, 291-301

Rock slide   45,106

Rocky Creek ash   42,85-89

Roman Wall   7,64

Roosevelt glacier   21,67,139-166

Schreibers Meadow ash, cinder cone, lava   8,10,21, 62,78,79,81-85,88,89,111, 136,203, 225,229,234

Sherman Peak   65,68-70,72,105-108,235,247

Sherman Crater   8,51,58,65,68-75,90-108

Sholes glacier   316-319

Squak glacier   82,221,222,235-246

Steam eruptions   74,90-104

Sulphur Creek lava   74,90-104

Swift Creek   10-13,15,17,19,46,48-50,89,111

Table Mt.   12,13,17,19,22,29-32,34,40-44,117, 133,135

Talum glacier   247-259

Thunder glacier   21,166,172

Twin Sisters Range   8,118,121,126,127,226

Twin Sisters dunite   121,126,127

Vancouver   6,7,90

Volcanic mudflows   109-113

Wells Creek   10,19,22,46

Younger Dryas   42,121,137,179,181,183, 203, 224,225,234,236,237,262

www.ingramcontent.com/pod-product-compliance
Lightning Source LLC
Chambersburg PA
CBHW081352290426
44110CB00018B/2357